T0324752

Springer Theses

Recognizing Outstanding Ph.D. Research

Aims and Scope

The series "Springer Theses" brings together a selection of the very best Ph.D. theses from around the world and across the physical sciences. Nominated and endorsed by two recognized specialists, each published volume has been selected for its scientific excellence and the high impact of its contents for the pertinent field of research. For greater accessibility to non-specialists, the published versions include an extended introduction, as well as a foreword by the student's supervisor explaining the special relevance of the work for the field. As a whole, the series will provide a valuable resource both for newcomers to the research fields described, and for other scientists seeking detailed background information on special questions. Finally, it provides an accredited documentation of the valuable contributions made by today's younger generation of scientists.

Theses are accepted into the series by invited nomination only and must fulfill all of the following criteria

- They must be written in good English.
- The topic should fall within the confines of Chemistry, Physics, Earth Sciences, Engineering and related interdisciplinary fields such as Materials, Nanoscience, Chemical Engineering, Complex Systems and Biophysics.
- The work reported in the thesis must represent a significant scientific advance.
- If the thesis includes previously published material, permission to reproduce this must be gained from the respective copyright holder.
- They must have been examined and passed during the 12 months prior to nomination.
- Each thesis should include a foreword by the supervisor outlining the significance of its content.
- The theses should have a clearly defined structure including an introduction accessible to scientists not expert in that particular field.

More information about this series at http://www.springer.com/series/8790

Gui Lu

Dynamic Wetting by Nanofluids

Doctoral Thesis accepted by
Tsinghua University, Beijing, China

 Springer

Author
Dr. Gui Lu
School of Energy Power and Mechanical
  Engineering
North China Electric Power University
Beijing
China

Supervisor
Prof. Yuan-Yuan Duan
Department of Thermal Engineering
Tsinghua University
Beijing
China

ISSN 2190-5053
Springer Theses
ISBN 978-3-662-48763-1
DOI 10.1007/978-3-662-48765-5

ISSN 2190-5061 (electronic)

ISBN 978-3-662-48765-5 (eBook)

Library of Congress Control Number: 2015953800

Springer Heidelberg New York Dordrecht London

Printed on acid-free paper

Springer-Verlag GmbH Berlin Heidelberg is part of Springer Science+Business Media
(www.springer.com)

Parts of this thesis have been published in the following journal articles:

- Lu G, Hu H, Duan YY, Sun Y (2013) Wetting kinetics of water nano-droplet containing non-surfactant nanoparticles: A molecular dynamics study, Appl Phys Lett, 103: 253104.
- Lu G, Duan YY, Wang XD (2014) Surface tension, viscosity, and rheology of water-based nanofluids: a microscopic interpretation on the molecular level. J Nanopart Res, 16: 2564
- Lu G, Duan YY, Wang XD (2015) Effects of free surface evaporation on water nanodroplet wetting kinetics: a molecular dynamics study. J Heat Transfer, 137: 091001-091001-6

Supervisor's Foreword

With the development of nanotechnology, we have entered a new era that the matter can be manipulated on an atomic or molecular scale. As one of the promising nanotechnology branches, nanofluids, fluids containing suspensions of nanoparticles, have been designed to enhance thermal properties and/or reduce drag coefficient for application ranging from heat exchangers to microfluidics. This makes them very attractive as working mediums in many applications over the past decades. The super-spreading by nanofluids, which was reported recently as a potential technique to "Spread the word about nanofluids," greatly extends the applications of nanofluids in microfluidic systems. The findings also provide a totally new approach to change the solid–liquid wettability from the aspects of fluids rather than the solids. Additionally, the super-spreading by nanofluids provides new material-manufacturing techniques using particle deposition processes to fabricate substrates for desired surface morphologies and properties. However, the study of nanofluid dynamic wetting encounters tremendous challenges, which lie not only in the lack of nanoscale experimental techniques to probe the complex nanoparticle random motion, sedimentation, or self-assembly, but also in the lack of multiscale or cross-scaled experimental technique or theory which describes the effects of nanoparticle motion occurring at the nanometer scale (10^{-9} m) on the dynamic wetting taking place at macroscopic scale (10^{-3} m).

The dissertation analyzes the complex dynamic wetting by nanofluids using sophisticated experimental methods and powerful multiscale simulation methods. The subject touches on a very broad area and different disciplines in physics (molecular, energy, thermodynamics), chemical engineering (surface and colloid), material science (nanomaterials), and also in computational science (large-scale parallel computing). The dissertation also provides a broad literature review on the recent developments in the areas of wetting kinetics of complex fluids and a first review on the research of nanofluid dynamic wetting. The dissertation intends to report the new phenomena and provide cross-disciplinary thinking, direction, and perspective on the still-fledgling field of nanofluid wetting kinetics. The findings in this book provide both multiscale (from nanoscale to macroscale) mechanisms and tunable methods to understand and control nanofluid dynamic wetting. Basic physical insights and fundamental understanding of the complex phenomena were provided, linking the fundamental science with practical and engineering applications.

As the supervisor of Dr. Lu, I am glad to recommend this dissertation to readers, particularly those specialized or interested in the surface and material science or thermal engineering, or university researchers, R&D engineers and graduate students who wish to have basic physical insights and fundamental understanding of the complex wetting kinetics phenomena.

Beijing   Prof. Yuan-Yuan Duan
July 2015

Acknowledgments

I realized that I have many friends to thank when I finished this book. However, I will not list all of them here because of the length of the list and the possibility of missing some of them. They all make my experience here incredibly enriching and rewarding. But there are four persons who are so special that I must acknowledge. I would not have been able to make the contribution to this book if it were not for my Ph.D. thesis advisor, Professor Yuan-Yuan Duan. I would like to express my most sincere gratitude to him for his continuous guidance, support, and encouragement throughout my doctoral study. His extensive knowledge, professionalism, ethic, and kindness will always be a source of inspiration for me. I am deeply grateful to Professor Xiao-Dong Wang at North China Electric Power University in Beijing. He showed me an example of excellence and dedication. I am also grateful to Professor Ying Sun at Drexel University in Philadelphia for the fruitful collaboration and everything I learned from her during my visiting study in USA. I am deeply grateful to the late Professor Xiao-Feng Peng, who hand-picked me when I was a freshmen in 2002 and continued to supervise me until September 10, 2009. Although he left us at an early age, his spirit will continue to inspire us.

I would like to dedicate this book to my wife, Rui-Xue Wang, for her love and understanding. She has always been supportive and cheered me up during difficult time. Most of my thanks go to my parents, my wife's parents, and my sisters, for their endless love, constant support, and encouragement over the years.

This work was financially supported by the National Natural Science Foundation of China (Grant Nos. 21176133 and 51321002).

Contents

Chapter 1
Introduction

Abstract This chapter describes the importance of dynamic wetting research and its application in modern industry. Some important physical concepts and principles involved in dynamic wetting phenomena have been reviewed. This chapter also reviews and summarizes the new achievements and contributions of recent investigations in the topics of dynamic wetting by complex fluids. The latest development on the research of nanofluid dynamic wetting is also described. In the last part of this chapter, the research roadmap of this book is provided.

1.1 Backgrounds

Dynamic wetting is a process that one fluid replaces the other fluid on the solid surfaces, which is a very common natural phenomenon. The dynamic wetting is widely involved in our daily activities and industrial applications. There are many examples of wetting in our daily life: the morning dew hanging on the grass, the rain drops rolling on the windows, the coffee staining the table, the water rising in the capillaries of a tree. The dynamic wetting also takes place in many industrial processes, such as coating, oil exploration, film manufacturing, printing, food production, dyeing, and mineral flotation. In thermal engineering, the flow and phase change of working mediums are two fundamental processes in many energy convert and thermal management devices. The dynamic wetting, which relates to the fluid flow and phase-change behaviors, plays significant roles in these energy utilization systems. Particularly, the dynamic wetting is very important in various microfluidic systems, such as the microchips or biochips, in which the surface tension plays significant roles when the system size reduces.

The contact line motion is the core problem that dynamic wetting deals with. The relationship between the dynamic contact angle and the contact line velocity (θ_D–U), as well as the relationship between the spreading radius and the spreading time (R–t), is usually used to describe the dynamic wetting of fluids on solid surfaces. These two relationships not only present the wettability of the fluids on the

© Springer-Verlag Berlin Heidelberg 2016
G. Lu, *Dynamic Wetting by Nanofluids*,
Springer Theses, DOI 10.1007/978-3-662-48765-5_1

Fig. 1.1 Schematic of
"contact line paradox"

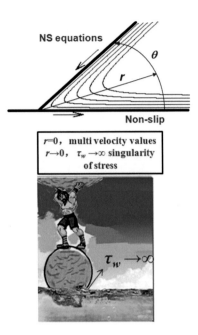

Fig. 1.1 Schematic of
"contact line paradox"

solid surfaces, but also show the energy dissipation mechanisms during the
dynamic wetting process [1–4]. The time-dependent dynamic wetting is of great
practical interest because fluids are usually used in flow devices. The dynamic
wetting process can be divided into two categories: the forced wetting and the
spontaneous wetting. In the forced wetting, the contact line motion is triggered by
the external forces, for example, the dynamic wetting of fluid in the capillary tube is
driven by the external pressure; the other example is the electrowetting, which is
driven by the external electricity field. In the spontaneous wetting, the dynamic
wetting takes place without any external forces (except the gravity). The droplet
spreads outwards to reduce the contact angle to reach the equilibrium stage with the
lowest systemic free energy, corresponding to the equilibrium contact angle.

The dynamic wetting process looks very simple. However, complexity always
hides beneath the simplicity. The complication of dynamic wetting process lies in
several facts: (1) The process is driven by multiple driven forces, such as viscosity
force, inertial force, gravity, and capillary force; (2) the process is affected by
various properties of spreading fluids, such as surface tension, viscosity, or rheology; (3) the process is also sensitive to various solid surface properties, such as
surface roughness, porosity, or surface charge; (4) the process usually occurs with
complex external physical fields, such as electricity field, magnetite field, or thermal
field. The complexity of dynamic wetting also lies in the famous "contact line
paradox" or "stress singularity" [2]. As shown in Fig. 1.1, if we solve the flows near
the contact line region, using the classical Navier–Stokes (NS) equations and the

non-slip boundary conditions, we will obtain an infinite force acting at the three-phase contact line. With such an infinite force, the contact line cannot move. However, the dynamic wetting process always takes place spontaneously in nature. Therefore, "not even Herakles could sink a solid if the physical model were entirely valid" [2]. The "contact line paradox" was initially proposed by Huh and Scriven in 1971 and drew extensively research interests on this topic. However, the mystery is still waiting for being unveiled up to now.

Recently, the researchers have paid widely attentions in nanofluid dynamic wetting. Adding nanoparticles into base fluids can greatly change the thermal conductivity, thermal diffusion coefficient, or phase-change behaviors. Therefore, nanofluids have been regarded as one of the promising heat transfer enhancement technologies. Nanofluids have been reported to increase the boiling heat transfer coefficient or critical heat flux (CHF) value, and hence preventing the boiling crisis. The improvement in the nanofluid boiling is not only attributed to the modification of thermal properties, but also to the dynamic wetting behaviors. The special dynamic wetting properties of nanofluids can provide us with a new approach to control the fluid wetting process, with which the solid–liquid wettability can be tuned from the aspect of fluids rather than solid surfaces. The advantage of the tuning the wettability using nanofluids lies in the multiformity of nanofluids, because there are many tunable parameters in nanofluids, such as nanoparticle material, shape, diameter, loading fraction, wettability, as well as base fluid. The tunable nanofluid dynamic wetting behaviors extend the potential applications of nanofluids to many scientific and engineering areas. For example, we can manipulate the nanoparticle self-assembly with the tunable nanofluid wettability. Then, we can fabricate the functional surfaces with desired physical, chemical, or optical characteristics, such as super-hydrophobic or super-hydrophilic surfaces, invisible coating on the fighter surfaces. Nanofluids are also widely involved in the bio/medicine engineering. The dynamic wetting plays a key role in the transport of bio/medicine nanoparticles in the biochips or other microfluidic devices.

For nanofluids, the additional nanoparticles induce the complex particle–particle and particle–solvent molecule interaction, which makes the dynamic wetting by nanofluids more complex. As shown in Fig. 1.2, the study of nanofluid dynamic wetting encounters two tremendous challenges. The first one is the lack of nanoscale experimental technique or theoretical description of complex nanoparticle random motion, sedimentation, or self-assembly. In addition, the dynamic wetting of nanofluids was a combination process governed by multiple forces. These forces cross several length scales, from 10^{-3} to 10^{-9} m. The lack of multiscale or cross-scaled experimental technique or theoretical description also prevents the well understanding of the mechanisms of nanofluid dynamic wetting.

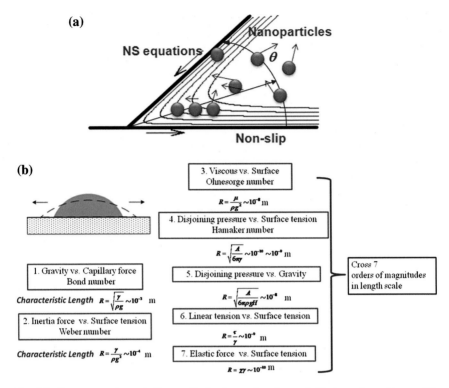

Fig. 1.2 Two tremendous challenges lie in the study of nanofluid dynamic wetting. **a** The nanoscale nanoparticle motions and the solid–liquid interaction. **b** Multiscale and cross-scaled characteristics

1.2 Literature Review

1.2.1 Dynamic Wetting of Newtonian Fluids

In order to reveal the complex dynamic wetting phenomenon, the early studies mainly focused on the simple fluids and ideal solid surfaces without considering the roles of any external physical fields. Extensive experiments [5–13], theories [14–42], and numerical simulations [43–62] have been conducted to reveal the fundamental mechanisms of the dynamic wetting by Newtonian fluids.

Among the experiments, several experimental techniques have been developed to measure the dynamic contact angle and the spreading radius during the dynamic wetting process, such as droplet spreading, Wilhelmy plate dipping, plunging tape, and capillary tube [5], which are summarized in Table 1.1.

Among the theoretical works, the dynamic wetting of Newtonian fluids has been studied extensively in last five decades. The motivations of the early study mainly focused on the "contact line paradox" issue or the physical mechanism of contact

Table 1.1 Experimental techniques in the dynamic wetting

Methods	Schematic	Wetting modes	Parameters	Fundamental
Droplet spreading		Spontaneously wetting	$R - t$; $\theta_D - U$	Optical method
Wilhelmy plate dipping		Forced wetting	$\theta_D - U$	Mechanical method
Plunging tape		Forced wetting	$\theta_D - U$	Optical method
Capillary tube		Forced wetting	$\theta_D - U$	Optical method

line movement. One general approach among these studies was using slip boundary condition rather than non-slip condition to remove the stress singularity. The paradox was solved by the math, but not by the physics. The physical connotation of contact line movement is still beyond the satisfactory explanation [15–20]. Another way to remove the stress singularity was using a hypothetical monomolecular thin-film layer ahead of the nominal contact line moving on the solid surface, which was known as precursor layer. The singularity point was removed to infinite far away by the precursor layer from the apparent contact line. Therefore, the "contact line paradox" can be avoided [21–23].

By introducing various microscopic hypotheses to remove the stress singularity at the contact line, several theoretical models have been established to describe the dynamic wetting process of Newtonian fluids. Among these models, the hydrodynamic model (HD) [15–23] and the molecular kinetic theory (MKT) [24] are most influential.

In HD, the "stress singularity" is assumed to only occur in the microregion near the contact line. In this region, continuous assumptions are no longer held due to the modification of fluid microscopic properties by the solid surface microstructure, the fluid heterogeneity, or the non-Newtonian effects. Therefore, the traditional NS equations with classical non-slip boundary conditions fail to describe the flows in this region. Additional microscopic assumptions, such as the precursor layer, slip boundary, or shear-thinning non-Newtonian conditions are needed to solve the flow fields. The evolution of droplet shape, dynamic contact angle, and the contact line moving velocity can be derived from the HD.

Another popular model is called MKT, which completely abandons the continuous assumptions in the NS equations and seeks a new approach to describe the contact line motion problems. In MKT, the principal hypothesis is that the motion of the three-phase line is ultimately determined by the statistical kinetics of molecular adsorption/desorption events occurring within the three-phase zone. For the contact line to advance, a forward-direction shear stress to the molecules within three-phase zone modifies the profiles of the potential energy barriers to molecular displacements in the forward direction. Therefore, the displacements in the forward direction are more frequent than those in the reverse direction, leading to the advancing of contact line.

Both of these models contain several microscopic parameters, such as the slip length, L_s, in the HD model; the molecular displacement frequency, K; and the displacement distance, λ, in the MKT model. These parameters are immeasurable, but can be fitted from the macroscopic dynamic wetting data, such as $\theta_D - U$ and $R - t$. Therefore, the experimental dynamic wetting data can provide us a tool to look inside the microscopic pictures in the microzone of contact line region.

The typical dynamic wetting models of Newtonian fluids are listed as follows.

1. Hydrodynamic model

   The HD can be divided into two categories: one is purely hydrodynamic approach (PHA) or standard hydrodynamic approach, which is strictly derived from the momentum equations; the other is energy-balanced approach (EBA), in which the dynamic wetting is considered as an energy dissipation process [25]. The typical HDs of Newtonian fluids were summarized in Table 1.2.

2. Molecular kinetic model

   Figure 1.3 illustrates the molecular displacement near the contact line region in the MKT model [24, 35, 36]. The absorbed sites locate randomly on the initial solid–liquid interface. The liquid molecular replacements occur randomly but progressively within the moving three-phase contact line zone. At equilibrium, the molecular displacement frequency in the forward direction equals to the frequency in the backward direction. However, when the contact angle diverges from the equilibrium value, the contact line movement is triggered by the unbalanced Young's stress, $F_Y = \sigma_{LV}(\cos\theta_0 - \cos\theta_D)$. Young's stress

Table 1.2 Typical hydrodynamic models of Newtonian fluids

Authors	Model	Equation	Description
Dussan [18]	PHA	$G(\theta_D) = G(\theta_{cl}) + \mathrm{Ca}\,\ln\left(\frac{L_H}{L_S}\right)$	Two zones
Cox [26]	PHA	$G(\theta_D) = G(\theta_{cl}) + \mathrm{Ca}\left[\ln\left(\frac{L_H}{L_S}\right) + \frac{Q_{in}}{f(\theta_{cl})} - \frac{Q_{out}}{f(\theta_D)}\right]$	Three zones
Zhou and Sheng [27, 28]	PHA	$G(\theta_D) = G(\theta_{cl}) + 9\mathrm{Ca}\left[\ln\left(\frac{L_H}{L_S}\right) + C_{ZS}\right]$	Slip length normalization
de Gennes [1, 31], Brochard-Wyart [32, 33]	EBA	$\theta_D(\theta_D^2 - \theta_0^2) = 6\mathrm{Ca}\,\ln\left(\frac{L_H}{x_m}\right)$	Young unbalanced force and viscous dissipation

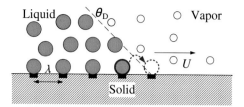

Fig. 1.3 Schematic of molecular displacement near the contact line region in MKT model [35]

modified the profiles of the potential energy barriers to molecular displacement, lowering the barriers in the forward direction and raising those in the backward direction. Consequently, the molecular displacements in the forward direction become more frequent than those in the reverse direction, leading to the contact line motion in the macroscopic scale. The contact line velocity is related to the displacement frequency K and the average distance between the two adsorption sites, λ, which gives $U = K\lambda$. Therefore, we have the relation of the dynamic contact angle and the contact line velocity,

$$\cos \theta_D = \cos \theta_0 - \frac{2k_B T}{\sigma_{LV} \lambda^2} \text{arcsinh} \left(\frac{U}{2K\lambda} \right), \tag{1.1}$$

in which θ_0 is the equilibrium contact angle, θ_D is the dynamic contact angle, k_B is Boltzmann constant, σ_{LV} is the liquid–vapor surface tension, K is the molecular displacement frequency, λ is the average distance of molecular displacement.

The advantage of MKT lies in its microscopic instinct in the explanation of contact line motion, without the help of slip hypothesis. In addition, the solid surface characteristics can be considered in the model. However, the model does not consider the viscous effects. The two models provide two explanations from different aspects to the contact line motion. In the HD model, the energy dissipation during the dynamic wetting occurs in the bulk droplet, referred to as the bulk dissipation or the viscous dissipation. In this model, the apparent contact angle is defined by either the outer region angle or the included angle that the moving meniscus extends to the solid–liquid interface. However, in the MKT model, the energy dissipation taking place in the vicinity of contact line zone dominates the spreading energy dissipation process, which is known as the local dissipation. The moving meniscus is determined by Young–Laplace equation. Therefore, the macroscopic contact angle in the MKT is strictly defined. It should be noted that there is still controversy whether λ depending on the liquid properties or solid surface properties in MKT [8, 38, 39]. Blake, who initially proposed the theory, claimed that the parameter λ might depend on the molecular size, but is determined, to a much greater extents, by the distance of the absorption sites on the solid surface [36]. The MKT model has been widely applied in the Newtonian fluid dynamic wetting process [36–40].

Table 1.3 Comparison of two dynamic wetting models

	Hydrodynamic model	MKT model
Physical scenario		
Fundamental theory	Lubrication approximation	Eyring theory
Parameters	p, U, Ca, θ, L_H	λ, K, U, θ, G
Equations	$\theta_D \left(\theta_D^2 - \theta_0^2 \right) = 6\mathrm{Ca} \ln \left(\frac{L_H}{x_m} \right)$	$\cos \theta_D = \cos \theta_0 - \frac{2k_B T}{\sigma_{LV} \lambda^2} \operatorname{arcsinh} \left(\frac{U}{2K\lambda} \right)$
Microscopic parameters	x_m, L_s	λ, K
Dissipation mode	Viscous dissipation	Local dissipation
Scope of application	Small contact angle	Large contact angle

The two typical models provide the understanding of contact line motion from different aspects, and also provide good agreement with most experimental data. However, both models have their limitations in the scope of application. In addition, they both contain unmeasured microscopic parameters. Table 1.3 compares the two typical dynamic wetting models.

The spreading law, the relation of spreading radius versus spreading time, is another research topic in the dynamic wetting. The spreading law has been established for the complete wetting of Newtonian fluids. The spreading radius is expressed as the power law of spreading time, $R(t)$–at^α, in which α is the spreading exponent. The exponent is used to characterize not only the contact line velocity, but also the energy dissipation mode during the contact line moving. In HD models, in which the viscous dissipation dominates, $\alpha = 1/10$ for the capillary regime, while $\alpha = 1/8$ [22] for the gravitational regime. However, $\alpha = 1/7$ for the MKT model [15], in which the local dissipation dominates. The unique spreading law fails in the partial wetting cases. de Ruijter et al. [10] suggested that the partial wetting process can be divided into three stages: the early spreading stage with $R(t)$–$R_0 + at$, the middle stage with $R(t)$–$t^{0.1}$, and the last stage when the dynamic contact angle approaching the equilibrium contact angle, which gives $\Delta R(t)$–$\exp(-t/T)$, in which $\Delta R(t) = R_{eq} - R(t)$, T is a constant. von Barh et al. [42] used high-speed photography to study the partially wetting process. They found that the spreading laws depend on the fluid viscosity. The low-viscosity fluids spread over tens of milliseconds to reach equilibrium stage. The spreading radius approximately satisfies $R(t)$–$t^{0.5}$, however, the high viscosity fluids spreading with $R(t)$–$t^{0.1}$.

According to different time and length scale, there are five numerical simulation techniques, as shown in Fig. 1.4, the Ab initio method, the hybrid method of quantum molecular dynamic simulations and the classical molecular dynamics simulations, the classical molecular dynamics simulations (MD), the lattice

Fig. 1.4 Multiscale numerical simulation techniques

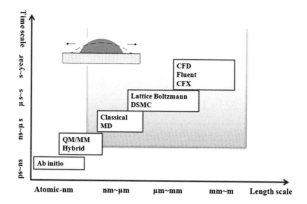

Boltzmann method (LBM), and the traditional Computational Fluid Dynamics (CFD) method. We do not need to consider the quantum effects during the dynamic wetting process, because, to some extent, wetting is a macroscopic process. However, the distinct drawback of CFD is significant when dealing with the dynamic wetting problems. On one hand, the evolutions of spreading radius and dynamic contact angle are inputted as boundary conditions in the CFD models when dealing with the wetting problems. Therefore, we could not study the dynamic wetting behaviors from CFD simulations. In addition, the non-slip boundary conditions used in the CFD methods are incapable to model the dynamic wetting process due to the "contact line paradox" as discussed in Sect 1.1 [2]. Molecular dynamics (MD) simulations have been recognized as a powerful tool in studying the contact line motion problems. However, most MD simulations focused on simple Lennard-Jones (LJ) fluid droplet spreading on LJ substrates, a system that is quite different from real dynamic wetting system and is hence impossible to mimic important physical properties (e.g., surface tension, density, and viscosity) of fluids which is related to the dynamic wetting [43–54]. The LBM is based on mesoscopic kinetic equations (the Boltzmann equation). The NS equations can be derived from the Boltzmann equation using the Chapman–Enskog multiscale expansion. Therefore, the LBM has been regarded as a very promising method to simulate the multiscale problems. Recently, a number of studies have used the LBM to analyze the flow and heat transfer of nanofluids using a single-component single-phase model [26–34] or a multicomponent single-phase model [35–38]. Due to the intrinsic microscopic kinetics, the LBM was widely used to investigate dynamic wetting [55–62]. However, most of these studies focused on the simple Newtonian fluid dynamic wetting.

In conclusion, the Newtonian dynamic wetting has been studied for almost 5 decades. Although the physical consensus is still unreached on the mechanism of contact line motion, a large amount of experimental data has been reported. Several theoretical models have been proposed to describe the relations of θ_D–U and R–t. Microscopic and mesoscopic scale simulation techniques have been established to study the dynamic wetting of Newtonian fluids.

1.2.2 Dynamic Wetting with Complex Surfaces and Complex External Physical Fields

Due to the complexity of dynamic wetting, the early studies usually focused on the dynamic wetting of simple fluids on an ideal smooth surface without any external physical fields. These studies can provide the fundamental understanding of dynamic wetting process, satisfying the curiosity of exploring the unknown word. However, it is hard to find simple fluids and ideal surfaces in the real word. Therefore, it is of greater practical interest to study the dynamic wetting of complex fluids on the real solid surface with external physical fields.

The gold of studying dynamic wetting is to tune the wettability or dynamic wetting behaviors of fluids on solid surfaces. Usually, tuning the solid surface properties is the most favorite option which draws extensive research interesting. For a liquid drop on the smooth and ideal surfaces, the static contact angle is a single value, which can be determined by the Young's equation, $\cos\theta_Y = (\sigma_{SV} - \sigma_{SL})/\sigma_{LV}$. However, for the real solid surfaces, a drop placed on a surface has a spectrum of contact angles ranging from the so-called advancing (maximal) contact angle, θ_A, to the so-called receding (minimal) contact angle, θ_R. The difference between these values, $\theta_A - \theta_R$, is known as contact angle hysteresis, which is usually attributed to the roughness or the heterogeneity of the solid surface [63–66]. The contact angle hysteresis is the fundamental guideline for the designs of various lotus biomimetic or nanostructured surfaces, with super-hydrophobic or super-hydrophilic characteristics [67–70], as shown in Fig. 1.5.

Other complex facts lie in that the dynamic wetting process usually takes place with various external physical fields, such as electric, magnetic, thermal, or optical fields [72–77]. The electrowetting [72], which is the modification of the wetting properties of a surface (which is typically hydrophobic) with an applied electric field, is one of the hot topics in this area. A specific example for the dynamic wetting with complex external fields is the liquid water transport within the gas diffusion layer in the proton exchange membrane (PEM) fuel cells, in which the dynamic wetting of liquid water occurs with the external electric field, electro-osmosis, and temperature gradient, as well as the complex porous structures. By tuning the wettability of gas diffusion layer, the output power of PEM fuel cells can be improved significantly [78].

1.2.3 Dynamic Wetting of Complex Fluids

The viscosity of Newtonian fluids is independent of the shear stress rate. However, for most of the fluids, especially for most of the mixture fluids, the viscosity depends on shear rate or shear rate history and exhibits non-Newtonian characteristics. Many polymer solutions and particulate suspensions are non-Newtonian fluids, as are many commonly found substances such as ketchup, custard,

Fig. 1.5 Dynamic wetting on the complex solid surfaces. **a** Lotus leaf effects. **b** Rose leaf effects. **c** Wettability on hybrid micro-nanostructural surfaces [71]

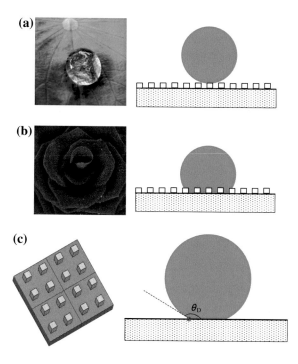

toothpaste, starch suspensions, paint, blood, and shampoo. There is extensive research of dynamic wetting with Newtonian fluids in the literature, but only a few theoretical and experimental studies have explored how non-Newtonian fluids spread over solid substrates. It is of greater practical interest to study the non-Newtonian fluid dynamic wetting. However, the study of non-Newtonian fluid dynamic wetting had just begun in the last two decades [71]. The current awareness of the non-Newtonian fluid dynamic wetting is far less than that of Newtonian fluid, which can be attributed to the complex nonlinear constitutive relations, or the diversity properties in non-Newtonian fluids.

The previous studies on non-Newtonian fluid dynamics wetting mainly focused on the relatively simple power-law fluids (shear-thinning and shear-thickening fluids). Carré and Woehl [79, 80] reported the dynamic wetting behaviors of shear-thinning fluids and derived a $\theta_D - U$ relation based on the "energy-balanced method" proposed by de Gennes's [1]. They found that the $\theta_D - U$ strongly depends on the rheological index, n. However, Neogi and Ybarra [81] reported a contrary result that the $\theta_D - U$ relation was independent of the fluid rheology. They also used the energy-balanced method to build the $\theta_D - U$ relations of Ellis and Reiner-Rivlin fluids. The viscous dissipation was analyzed using approximative method because the constitutive relationships of these two fluids are of complexity. The authors suggested that the non-Newtonian fluid dynamic wetting can be described by Newtonian fluid $\theta_D - U$ relation, only changing the viscosity of the relation into the non-Newtonian fluid viscosity at zero shear rate.

Starov et al. [82] analyzed the complete wetting of power-law non-Newtonian fluids using the lubrication theory. Their results showed that the dynamic contact angle of the power-law fluid droplet was related to the droplet sizes during the dynamic wetting process, which was quite different from Newtonian fluids. In addition, for both the capillary and gravitational regimes, the spreading exponents of Newtonian fluids were higher than that of shear-thinning fluids, but lower than shear-thickening fluids.

Betelu and Fontelos [83, 84] also used the lubrication approximate to study the complete spreading of power-law droplet. They only considered the shear-thinning fluids in the capillary regime. The shape evolution of free surface was solved numerically using similarity transformation. Wang et al. [85, 86] established a two-dimensional model to describe the power-law fluid droplet spreading by solving the ordinary differential equations of film thickness using traveling wave transformation. Liang et al. [87, 88] established two power-law fluid dynamic wetting models based on the HD and MKT, which all agree well with the experimental data. Table 1.4 compares various non-Newtonian dynamic wetting models.

Compared with the Newtonian fluids, very few experimental data of the non-Newtonian fluid dynamic wetting have been reported. In addition, most of the experiments focused on the shear-thinning fluids, as shown in Table 1.5. Carré and Woehl [79] tested the shear-thinning fluid droplets (PDMS+silica and acrylic typographic ink) spreading on the glass slides, which agreed with their non-Newtonian fluid dynamic wetting model. Rafai and Bonn [89, 90] studied the shear-thinning (xanthan solution) and normal stress fluid (polyacrylamide solution) dynamic wetting on mica using the droplet spreading method. Their results show that the spreading exponents of the two types of fluids in their experiments are both less than 1/10, the spreading exponent of Newtonian fluids [91]. In addition, the exponent decreases with increasing solution concentrations due to the more prominent non-Newtonian behaviors. The results verified the Starov's finding that the spreading exponents of shear-thinning fluids should be less than that of Newtonian fluids. Wang and Duan's group [85, 86] have done many relevant works

Table 1.4 Theoretical studies of non-Newtonian dynamic wettings

Authors	Types of non-Newtonian fluids	Wettability	Models
Carré and Woehl [79, 80]	Shear thinning	Completely/partially	Hydrodynamics
Starov et al. [82]	Shear thinning/shear thickening	Completely	Hydrodynamics
Betelu et al. [83, 84]	Shear thinning	Completely/partially	Hydrodynamics
Wang et al. [85, 86]	Shear thinning	Completely/partially	Hydrodynamics
Liang et al. [87, 88]	Shear thinning/shear thickening	Completely/partially	Hydrodynamics/MKT

Table 1.5 Experiments on the non-Newtonian fluid dynamic wetting

Authors	Fluid types	Fluids/substrates	Methods
Carré and Woehl [79]	Power law	SiO_2-PDMS/glass	Droplet spreading
Rafai et al. [89, 90]	Power law	Xanthan/mica	Droplet spreading
Wang et al. [85, 86]	Shear thinning/shear thickening	Multiple fluids/multiple surfaces	Droplet spreading
Digilov [93]	Power law	Xanthan/capillary tube	Capillary rise
Min et al. [92]	Shear thinning	Multiple fluids/multiple surfaces	Droplet spreading/Wilhelmy plate dipping

on the non-Newtonian fluid dynamic wetting: they first studied the dynamic wetting behaviors using two experimental techniques, droplet spreading method and Wilhelmy plate method, corresponding to the spontaneous wetting and forced wetting, respectively; they extended the scope of complex fluids from power-law fluids to more complex fluids without general constructive relations; they proposed several theoretical models with pure HD, energy-balanced model, and the molecular kinetic theory; they also conducted multiscale simulation techniques, the mesoscopic LBM, and the microscopic MD simulations, to study the dynamic wetting behaviors of complex fluids. It is noted that they contributed several significant findings in the field of non-Newtonian fluid dynamic wetting. For example, by introducing a new defined capillary number (Ca), they integrated the present diverse non-Newtonian fluid dynamic wetting models into a general model; they also found that the dynamic wetting of non-Newtonian fluids strongly depends on the macrogeometry, which is quite different from the Newtonian fluids [92].

1.2.4 Studies of Dynamic Wetting by Nanofluids

Nanofluids, fluids containing suspension of nanometer-sized particles, can also be regarded as one type of complex fluids. Nanofluids are promising branches in nanotechnologies which manipulate matter with at least one dimension sized from 1 to 100 nm. Nanotechnology is able to create many new materials and devices with a vast range of applications, such as interface and colloid science, nanoscale materials, nanomedicine, nanoparticles, and biomedical applications. As part of this technology, nanofluids have also been widespread concerned. The concept of nanofluid was first proposed by Choi [94]. The nanofluids are designed to enhance thermal properties and/or reduce drag coefficients for application ranging from electronics cooling to microfluidics. The suspended nanoparticles can significantly

Table 1.6 Critical heat flux (CHF) in the nanofluid boiling

Authors	Nanofluids	CHF enhancement (%)
You et al. [111]	Al_2O_3/water	200
Kim et al. [112]	TiO_2/water	200
Vassallo et al. [113]	SiO_2/water	60
Tu et al. [114]	Al_2O_3/water	67
Kim and Kim [115]	TiO_2/water	50
Moreno et al. [116]	Al_2O_3/water, ZnO/water; Al_2O_3/silicone oil	200
Bang et al. [117]	Al_2O_3/water	50
Milanova et al. [118]	SiO_2/water, CeO/water, Al_2O_3/water	170
Jackson et al. [119]	Au/water	175
Wen and Ding [120]	Al_2O_3/water	40

modify the transport properties of the base fluids, and the resulting nanofluids exhibit attractive properties such as high thermal conductivity and high boiling heat transfer coefficient [95–102]. For example, with 1 % nanoparticle loading, the thermal conductivity of base fluid can be increased 40 % [100].

Recently, it is reported that nanofluids exhibit enhanced or tunable dynamic wetting characteristics compared with the base fluids [103–110]. The tunable wetting behaviors can be used to extend the applications of nanofluids into many areas such as fabricating functional surfaces with desired properties, enhancing heat transfer or reducing the drag in micro/biofluid systems, or enhancing boiling or condensation phase change in heat exchangers. In addition, the dynamic wetting by nanofluids also offers a new approach to tune the wettability of fluids on the solid surfaces, which is based on the fluids rather than the solid surfaces. This approach provides more options for us to tune the wettability of fluids on solid surfaces, because there are many tunable parameters in nanofluids, such as nanoparticle loading, material, size, shape, as well as the base fluid material. Thus, nanofluids may be smart materials if their wettability can be manipulated. Therefore, the dynamic wetting by nanofluids has become a hot topic in many areas, from thermal science, material science, to the colloidal and interfacial science.

The dynamic wetting by nanofluids attracted research attentions firstly by some interesting phenomena: Adding nanoparticles into base fluids can modify the equilibrium contact angle, change the surface tension, or result in the solid-like ordering structure of nanoparticles assembled near the contact line; anomalous evaporating or boiling behaviors by nanofluids [103, 107, 111–120]. These phenomena were found to be related to the different dynamic wetting behavior of nanofluids compared with their base fluids. One motivation of studying nanofluid dynamic wetting in the thermal engineering community lies in the enhancement of the heat transfer coefficient and the CHF [111–120] during the boiling process, as shown in Table 1.6.

The enhancement in the liquid–vapor phase change was explained by the improvement of nanofluid dynamic wetting. Thus, the abnormal dynamic wetting behavior of nanofluids, especially the contact line motion affected by the addition nanoparticles, stimulates the research of nanofluid dynamic wetting. Through years of exploration work, researchers have been basically reached these consensuses: the dynamic wetting itself is a complex process; adding nanoparticles into base fluids will lead to more complex dynamic wetting by inducing the complicated nanoparticle random motion in the bulk liquid, or self-assembly at the liquid–vapor interface or in the solid–liquid–vapor three-phase contact line region; the process will be more unpredictable if the dynamic wetting occurs with phase change or complex external fields. The mechanism of nanofluid dynamic wetting is still unclear due to the lack of macroscopic and microscopic experiments and theories.

There are several explanations for the roles of adding nanoparticles in the dynamic wetting: (1) Adding nanoparticles into the base fluids was reported to change the dynamic wetting behaviors by modifying the rheological properties of nanofluids. The dynamic wetting is enhanced by shear-thinning nanofluids but is hindered by shear-thickening nanofluids [13]. (2) The heterogeneity due to the nanoparticle self-assembly also affects the dynamic wetting by nanofluids. Nanofluid "super-spreading" was reported by Wasan et al. [107] who found that an 8 nm micellar solution, an 1 μm latex suspension, and a 20 nm silica suspension were found to enhance the base fluids wettability [106–108]. A solid-like ordering structure of nanoparticles was observed near the contact line region using inter-ferometry. This solid-like ordering structure stemming from the settlement and assembly of nanoparticles gives rise to a structural disjoining pressure in the vicinity of the contact line. This excess pressure in turn alters the force balance near the contact line and enhances the spreading of nanofluids. The super-spreading behavior of nanofluids induced by the self-assembly of the nanoparticles and the structural disjoining pressure have been widely used to explain the enhanced dropwise evaporation [18, 19] and CHFs with nanofluids [20–23]. (3) Another explanation for the super-spreading by nanofluids was that nanoparticles were assumed to settle at the bottom of the droplet; thus, reducing the solid–liquid friction and, hence, facilitating the fluid spreading [103]. The last two explanations are schematically shown in Fig. 1.6. However, these two explanations are only qualitative assumptions; both adequate experimental evidences and theoretical descriptions are still needed.

1.3 Objectives of the Dissertation

According to the literature review, the mechanism of dynamic wetting by nano-fluids is still unclear due to limitations of nanoscale experimental techniques and fundamental theories. Studies of the dynamic wetting by nanofluids are facing great challenges because the wetting behavior crosses several length and timescales. This study analyzes the effects of the bulk and local dissipation in the nanofluids due to

Fig. 1.6 Schematics of
super-spreading due to
nanoparticle self-assembly.
a Self-assemble at the contact
line region. **b** Bottom
sedimentation

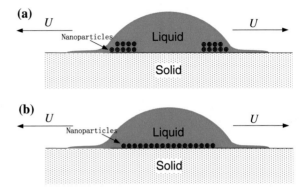

the nanoparticle transport and self-assembly on the macroscopic dynamic wetting
behavior using macroscopic experiments and multiscale simulation methods. The
results describe both the macroscopic and microscopic mechanisms and tunable
methods to control nanofluid dynamic wetting. The research road map is shown in
Fig. 1.7.

In Chap. 2, the contact line mobility and contact angle evolution were first
measured using the droplet spreading method and the Wilhelmy plate method. The
effects of the nanofluid parameters, such as nanoparticle loading, particle diameter,
particle material, and base fluid, were examined, as well as the effects of the
substrate material.

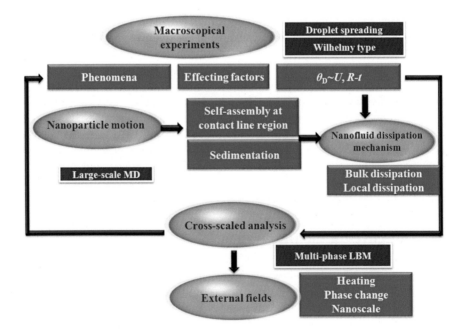

Fig. 1.7 Schematic of research road map

In Chaps. 3 and 4, MD simulations were then conducted to study the nanoparticle transport and self-assembly within the bulk liquid or at the interface, which affect the bulk dissipation and the local dissipation. The results describe the microscopic mechanisms by which the bulk dissipation affects the nanofluid surface tension, viscosity, and rheology. Without nanoparticle self-assembly at the contact line, nanofluid dynamic wetting was controlled by the increasing surface tension and the increasing solid–liquid friction coefficient. MD simulations of nano-thin-film motion quantitatively illustrate how the capillary force and the "structural disjoining pressure" drive the contact line motion.

In Chap. 5, a mesoscopic understanding of the nanofluid wetting kinetics was then obtained using the LBM. The objective of this chapter was examining the effects of two nanoparticle dissipations, the bulk dissipation and the local dissipation, on the macroscopic dynamic wetting process. The roles of surface tension, rheology, and the structural disjoining pressure were investigated.

In Chap. 6, the nanofluid dynamic wetting was studied with complex external conditions. The effects of substrate heating and intensive free surface evaporation on the wetting kinetics of nanofluids were simulated using MD simulations. The effects of initial droplet temperature, substrate temperature, and wettability were examined. The microscopic mechanisms of nanoparticle self-assembly and nanofluid droplet evaporating–spreading behavior were revealed by traced the particle motion and molecular mobility near the contact line region.

Finally, in Chap. 7, we provided the general conclusions and the contributions of this book. The prospects in dynamic wetting by nanofluids were also provided in this chapter.

References

1. de Gennes PG (1985) Wetting: statics and dynamics. Rev Mod Phys 57:827–863
2. Huh C, Scriven LE (1971) Hydrodynamic model of steady movement of a solid-liquid-fluid apparent contact line. J Colloid Interface Sci 35:85–101
3. Dussan VEB (1979) On the shredding of liquids on solid surfaces: static and dynamic contact lines. Annu Rev Fluid Mech 11:317–400
4. Oron A, Davis SH, Bankoff SG (1997) Long-scale evolution of thin liquid films. Rev Mod Phys 69:931–980
5. Kistler SF (1993) Hydrodynamics of wetting. In: Berg JC (ed) Wettability. Marcel Dekker, New York, pp 311–429
6. Petrov PG, Petrov JG (1992) A combined molecular-hydrodynamic approach to wetting kinetics. Langmuir 8:1762–1767
7. Hayes RA, Ralston J (1993) Forced liquid movement on low-energy surfaces. J Colloid Interface Sci 159:429–438
8. Hayes RA, Ralston J (1994) The molecular-kinetic theory of wetting. Langmuir 10:340–342
9. de Ruijter MJ, De Coninck J, Blake TD (1997) Contact angle relaxation during the spreading of partially wetting drops. Langmuir 13:7293–7298
10. de Ruijter MJ, de Coninck J, Oshanin G (1999) Droplet spreading: partial wetting regime revisited. Langmuir 15:2209–2216

11. Schneemilch M, Hayes RA, Petrov JG (1998) Dynamic wetting and dewetting of a low-energy surface by pure liquids. Langmuir 14:7047–7051
12. Blake TD, Bracke M, Shikhmurzaev YD (1999) Experimental evidence of nonlocal hydrodynamic influence on the dynamic contact angle. Phys Fluids 11:1995–2007
13. de Ruijter MJ, Charlot M, Voue M (2000) Experimental evidence of several time scales in drop spreading. Langmuir 16:2363–2368
14. Blake TD, Shikhmurzaev YD (2002) Dynamic wetting by liquids of different viscosity. J Colloid Interface Sci 253:196–202
15. Hocking LM (1976) A moving fluid interface on a rough surface. J Fluid Mech 76:801–817
16. Huh C, Mason SG (1977) Steady movement of a liquid meniscus in a capillary tube. J Fluid Mech 81:401–419
17. Hocking LM (1977) A moving fluid interface, Part II: the removal of the force singularity by a slip flow. J Fluid Mech 79:209–229
18. Dussan VEB (1976) The moving contact line: the slip boundary conditions. J Fluid Mech 76:665–684
19. Lowndes J (1980) The numerical simulation of the steady movement of fluid meniscus in a capillary tube. J Fluid Mech 101:631–646
20. Bach P, Hassager O (1985) An algorithm for the use of the Lagrangian specification in Newtonian fluid mechanics and applications to free-surface flow. J Fluid Mech 152:173–190
21. Ludviksson V, Lightfoot EN (1968) Deformation of advancing menisci. AIChE J 14:674–677
22. Tanner LH (1979) The spreading of silicone oil drops on horizontal surfaces. J Phys D 12:1473–1484
23. Hervet H, de Genees PG (1984) The dynamics of wetting: precursor films in the wetting of dry solids. C R Acad Sci 299:499–503
24. Blake TD (1993) Dynamic contact angles and wetting kinetics. In: Berg JC (ed) Wettability. Marcel Dekker, New York, pp 251–309
25. Daniel RC, Berg JC (2006) Spreading on and penetration into thin, permeable print media: application to ink-jet printing. Adv Colloid Interface Sci 123:439–469
26. Cox RG (1986) The dynamics of the spreading of liquids on a solid surface I. Viscous flow. J Fluid Mech 168:169–194
27. Zhou MY, Sheng P (1990) Dynamics of immiscible-fluid displacement in a capillary tube. Phys Rev Lett 64:882–885
28. Sheng P, Zhou MY (1992) Immiscible-fluid displacement: contact-line dynamics and the velocity-dependent capillary pressure. Phys Rev A 45:5694–5708
29. Petrov JG, Ralston J, Schneemilch M et al (2003) Dynamics of partial wetting and dewetting in well-defined systems. J Phys Chem B 107:1634–1645
30. Voinov OV (1976) Hydrodynamics of wetting. Fluid Dyn 11:714–721
31. de Gennes PG (1986) Deposition of Langmuir-Blodgett layers. Colloid Polym Sci 264:463–465
32. Brochard-Wyart F, de Gennes PG (1994) Spreading of a drop between a solid and a viscous polymer. Langmuir 10:2440–2443
33. Brochard-Wyart F, Debrégeas G, de Gennes PG (1996) Spreading of viscous droplets on a non viscous liquid. Colloid Polym Sci 274:70–72
34. White FM (1998) Fluid mechanics. Mcgraw-Hill College, New York
35. Blake TD, Haynes JM (1969) Kinetics of liquid/liquid displacement. J Colloid Interface Sci 30:421–423
36. Blake TD, De Coninck J (2002) The influence of solid-liquid interactions on dynamic wetting. Adv Colloid Interface Sci 96:21–36
37. Glasstone S, Laidler KJ, Eyring H (1941) The theory of rate processes. McGraw-Hill, New York
38. Vega MJ, Gouttierè C, Seveno D (2007) Experimental investigation of the link between static and dynamic wetting by forced wetting of nylon filament. Langmuir 23:10628–10634

39. Ray S, Sedev R, Priest C (2008) Influence of the work of adhesion on the dynamic wetting of chemically heterogeneous surfaces. Langmuir 24:13007–13012
40. de Ruijter M, Kölsch P, Voué M (1998) Effect of temperature on the dynamic contact angle. Colloid Surface: A 144:235–243
41. Wang X, Chen LQ, Bonaccurso E (2013) Dynamic wetting of hydrophobic polymers by aqueous surfactant and superspreader solutions. Langmuir 29:14855–14864
42. von Bahr M, Tiberg F, Yaminsky V (2001) Spreading dynamics of liquids and surfactant solutions on partially wettable hydrophobic substrates. Colloids Surf A 193:85–96
43. Koplik J, Banavar JR, Willemsen JF (1988) Molecular dynamics of poiseuille flow and moving apparent contact lines. Phys Rev Lett 60:1282–1285
44. Thompson PA, Robbins MO (1989) Simulations of ACL motion: slip and dynamic contact angle. Phys Rev Lett 63:766–769
45. Yang JX, Koplik J, Banavar JR (1991) Molecular dynamic of droplet spreading on a solid surface. Phys Rev Lett 67:3539–3542
46. Blake TD, Clarke A, de Coninck J et al (1997) Contact angle relaxation during droplet spreading: comparison between molecular kinetic theory and molecular dynamics. Langmuir 13:2164–2166
47. Liu H, Chakrabarti A (1999) Molecular dynamics study of adsorption and spreading of a polymer chain onto a flat surface. Polymer 40:7285–7293
48. Hwang CC, Ho JR, Lee RC (1999) Molecular dynamics of a liquid drop spreading in a corner formed by two planar substrates. Phys Rev E 60:5693–5698
49. Yaneva J, Milchev A, Binder K (2003) Dynamics of a spreading nanodroplet: a molecular dynamic simulation. Macromol Theory Simul 12:573–581
50. He G, Hadjiconstantinou NG (2003) A molecular view of Tanner's law: molecular dynamics simulations of droplet spreading. J Fluid Mech 497:123–132
51. Shen YY, Couzis A, Koplik J et al (2005) Molecular dynamics study of the influence of surfactant structure on surfactant-facilitated spreading of droplets on solid surfaces. Langmuir 26:12160–12170
52. Bertrand E, Blake TD, de Coninck J (2005) Spreading dynamics of chain-like monolayers: a molecular dynamics study. Langmuir 21:6628–6635
53. Kim HY, Qin Y, Fichthorn KA (2006) Molecular dynamics simulation of nanodroplet spreading enhanced by linear surfactants. J Chem Phys 125:174708
54. Seveno D, Dinter N, de Coninck J (2010) Wetting dynamics of drop spreading: new evidence for the microscopic validity of the molecular-kinetic theory. Langmuir 26:14642–14647
55. Yang YT, Lai FH (2011) Numerical study of flow and heat transfer characteristics of alumina-water nanofluids in a microchannel using the lattice Boltzmann method. Int Commun Heat Mass 38:607–614
56. Kumar S, Prasad SK, Banerjee J (2010) Analysis of flow and thermal field in nanofluid using a single phase thermal dispersion model. Appl Math Model 34:573–592
57. Nemati H, Farhadi M, Sedighi K et al (2010) Lattice Boltzmann simulation of nanofluid in lid-driven cavity. Int Commun Heat Mass 37:1528–1534
58. Xuan YM, Li QA (2009) Energy transport mechanisms in nanofluids and its applications. In: 7th International conference on nanochannels and minichannels. Pohang, South Korea
59. Xuan YM, Yu K, Li Q (2005) Investigation on flow and heat transfer of nanofluids by the thermal lattice Boltzmann model. Prog Comput Fluid Dy 5:13–19
60. Xuan YM, Yao ZP (2005) Lattice Boltzmann model for nanofluids. Heat Mass Transfer 41:199–205
61. Khiabani RH, Joshi Y, Aidun C (2007) Convective heat transfer in a channel in the presence of solid particles. In: 7th ASME/JSME thermal engineering and summer heat transfer conference. Vancouver, Canada
62. Khiabani RH, Joshi Y, Aidun C (2010) Heat transfer in microchannels with suspended solid particles: lattice-Boltzmann based computations. J Heat Transfer 041003-1-9
63. Wenzel RN (1936) Resistance of solid surfaces to wetting by water. Ind Eng Chem 28:988–994

64. Cassie ABD (1948) Contact angle hysteresis on heterogeneous surfaces. Discuss Faraday Soc 3:11–16
65. Joanny JF, de Gennes PG (1984) A model for contact angle hysteresis. J Chem Phys 81:552–562
66. Schwartz LW, Garoff S (1985) Contact angle hysteresis and the shape of the three phase line. J Colloid Interface Sci 106:422–437
67. Quéré D (2005) Non-sticking drops. Rep Prog Phys 68:2495–2532
68. Dietrich S, Popescu MN, Rauscher M (2005) Wetting on structured substrates. J Phys-Condens Mat 17:S577–S593
69. Brusch L, Kuhne H, Thiele U et al (2002) Dewetting of thin films on heterogeneous substrates: pinning versus coarsening. Phys Rev E 66:011602
70. Darhuber AA, Troian SM, Reisner WW (2001) Dynamics of capillary spreading along hydrophilic microstripes. Phys Rev E 64:031603
71. Bonn D, Eggers J, Indekeu J et al (2009) Wetting and spreading. Rev Mod Phys 81:739–805
72. Quilliet C, Berge B (2001) Electrowetting: a recent outbreak. Curr Opin Colloid Interface Sci 6:34–39
73. Sondag-Huethorst JAM, Fokkink LGJ (1994) Potential-dependent wetting of electroactive ferricene-terminated alkanethiolate monolayers on gold. Langmuir 10:4380–4387
74. Verheijen HJJ, Prins MWJ (1999) Contact angle and wetting velocity measured electrically. Rev Sci Instrum 70:3668–3673
75. Blake TD, Clarke A, Stattersfield EH (2000) An investigation of electrostatic assist in dynamic wetting. Langmuir 16:2928–2935
76. Pollack MG, Fair RB, Shenderov A (2000) Electrowetting-based actuation of liquid microdroplets for microfluidic applications. Appl Phys Lett 77:1725–1726
77. Vallet M, Vallade M, Berge B (1999) Limiting phenomena for the spreading of water on polymer films by electrowetting. Eur Phys J B 11:583–591
78. Pai YH, Ke JH, Huang HF (2006) CF4 plasma treatment for preparing gas diffusion layers in membrane electrode assemblies. J Power Sources 161:275–281
79. Carré A, Woehl P (2002) Hydrodynamic behavior at the triple line of spreading liquids and the divergence problem. Langmuir 18:3600–3603
80. Carré A, Woehl P (2006) Spreading of silicon oils on glass in two geometries. Langmuir 22:134–139
81. Neogi P, Ybarra RM (2001) The absence of a rheological effect on the spreading of small drops. J Chem Phys 115:7811–7813
82. Starov VM, Tyatyushkin AN, Velarde MG et al (2003) Spreading of non-Newtonian liquids over solid substrates. J Colloid Interface Sci 257:284–290
83. Betelu SI, Fontelos MA (2003) Capillarity driven spreading of power-law fluids. Appl Math Lett 16:1315–1320
84. Betelu SI, Fontelos MA (2004) Capillarity driven spreading of circular drops of shear-thinning fluid. Math Comput Model 40:729–734
85. Wang XD, Lee DJ, Peng XF et al (2007) Spreading dynamics and dynamic contact angle of non-Newtonian fluids. Langmuir 23:8042–8047
86. Wang XD, Zhang Y, Lee DJ et al (2007) Spreading of completely wetting or partially wetting power-law fluid on solid surface. Langmuir 23:9258–9262
87. Liang ZP, Wang XD, Lee DJ et al (2009) Spreading dynamics of power-law fluid droplets. J Phys: Condens Matter 21:464117
88. Liang ZP, Wang XD, Duan YY et al (2010) Dynamic wetting of non-Newtonian fluids: multicomponent molecular-kinetic approach. Langmuir 26:14594–14599
89. Rafai S, Bonn D, Boudaoud A (2004) Spreading of non-Newtonian fluids on hydrophilic surfaces. J Fluid Mech 513:77–85
90. Rafai S, Bonn D (2005) Spreading of non-Newtonian fluids and surfactant solutions on solid surfaces. Phys A 358:58–67
91. Hoffman RL (1975) Study of advancing interface 1. Interface shape in liquid-gas systems. J Colloid Interface Sci 50:228–241

92. Min Q, Duan YY, Wang XD et al (2010) Spreading of completely wetting, non-Newtonian fluids with non-power-law rheology. J Colloid Interface Sci 348:250–254
93. Digilov RM (2008) Capillary rise of a non-Newtonian power law liquid: Impact of the fluid rheology and dynamic contact angle. Langmuir 24:13663–13667
94. Choi SUS (1995) Enhancing thermal conductivity of fluids with nanoparticle. In: Wang HP, Siginer DA (eds) ASME, FED 231/MD-66. New York
95. Hu ZS, Dong JX (1998) Study on antiwear and reducing friction additive of nanometer titanium oxide. Wear 216:92–96
96. Hamilton RL, Crosser OK (1962) Thermal conductivity of heterogeneous two component system. Ind Eng Chem Fundam 1:187–191
97. Lee S, Choi SUS, Li S (1999) Measuring thermal conductivity of fluids containing oxide nanoparticles. J Heat Transfer 121:280–289
98. Nagasaka Y, Nagashima A (1981) Absolute measurement of the thermal conductivity of electrically conducting liquids by the transient hot-wire method. J Phys E: Sci Instrum 14:1435–1440
99. Wang XW, Xu XF, Choi SUS (1999) Thermal conductivity of nanoparticle-fluid mixture. J Thermophys Heat Transfer 13:474–480
100. Sarit KD, Nandy P, Peter T (2003) Temperature dependence of thermal conductivity enhancement for nanofluids. J Heat Transfer 125:567–574
101. Murshed SMS, Leong KC, Yang C (2008) Investigations of thermal conductivity and viscosity of nanofluids. Int J Therm Sci 47:560–568
102. Das SK, Putra N, Roetzel W (2003) Pool boiling characteristics of nano-fluids. Int J Heat Mass Transfer 46:851–862
103. Sefiane K, Skilling J, MacGillivary J (2008) Contact line motion and dynamic wetting of nanofluid solutions. Adv Colloid Interface Sci 138:101–120
104. Vafaei S, Borca-Tasciuc T, Podowski MZ et al (2006) Effect of nanoparticles on sessile droplet contact angle. Nanotechnology 17:2523–2527
105. Trokhymchuk A, Henderson D, Nikolov AD et al (2001) A simple calculation of structural and depletion forces for fluids/suspensions confined in a film. Langmuir 17:4940–4947
106. Chaudbury MJ (2003) Spread the word about nanofluids. Nature 423:131–132
107. Wasan DT, Nikolov AD (2003) Spreading of nanofluids on solids. Nature 423:156–159
108. Chengara A, Nikolov AD, Wasan DT et al (2006) Spreading of nanofluids driven by the structural disjoining pressure gradient. J Colloid Interface Sci 280:192–201
109. Choi CH, Kim CJ (2006) Large slip of aqueous liquid flow over a nanoengineered superhydrophobic surface. Phys Rev Lett 96:066001
110. Sefiane K, Bennacer R (2009) Nanofluids droplets evaporation kinetics and wetting dynamics on rough heated substrates. Adv Colloid Interface Sci 147–148:263–271
111. You SM, Kim J, Kim KH (2003) Effect of nanoparticles on critical heat flux of water in pool boiling heat transfer. Appl Phys Lett 83:3374–3376
112. Kim H, Kim J, Kim M (2006) Experimental study on CHF characteristics of water-TiO_2 nano-fluids. Nucl Eng Technol 38:619–623
113. Vassallo P, Kumar R, Amico SD (2004) Pool boiling heat transfer experiments in silica-water nano-fluids. Int J Heat Mass Transfer 47:407–441
114. Tu JP, Dinh N, Theofanous T (2004) An experimental study of nanofluid boiling heat transfer. In: Proceedings of sixth international symposium on heat transfer. Beijing, China
115. Kim HD, Kim MH (2009) Critical heat flux behavior in pool boiling of water-TiO_2 nano-fluids. In: Proceedings of fourth japan-korea symposium on nuclear thermal hydraulics and safety. Sapporo, Japan
116. Moreno M, Oldenburg S, You SM et al (2005) Pool boiling heat transfer of alumina-water, zinc oxide-water and alumina-water ethylene glycol nanofluids. In: Proceedings of HT 2005. San Francisco, California
117. Bang IC, Chang SC (2005) Boiling heat transfer performance and phenomena of Al_2O_3-water nano-fluids from a plain surface in a pool. Int J Heat Mass Transfer 48:2407–2419

118. Milanova D, Kumar R, Kuchibhatla S et al (2006) Heat transfer behavior of oxide nanoparticles in pool boiling experiment. In: Proceedings of the fourth international conference on nanochannels, microchannels and minichannels. Limerick, Ireland
119. Jackson JE, Borgmeyer BV, Wilson CA et al (2006) Characteristics of nucleate boiling with gold nanoparticles in water. In: Proceedings of IMECE-2006. Chicago
120. Wen D, Ding Y (2005) Experimental investigation into the pool boiling heat transfer of aqueous based alumina nanofluids. J Nanopart Res 7:265

Chapter 2
Experimental Study on the Nanofluid Dynamic Wetting

Abstract In this chapter, the time-dependent wetting radius and contact angle for various nanofluid droplets were measured to study the dynamic wetting behaviors of nanofluids. The experiment results show that the adding of nanoparticles inhibits the dynamic wetting of nanofluids as compared with base fluids. The reduced spreading rate can be attributed to the increase in either surface tension or viscosity due to adding nanoparticles into the base fluid. Once the effects of the surfaces tension and viscosity are both eliminated using the non-dimensional analysis, the wetting radius versus spreading time curves for all the nanofluid droplets overlap with each other. The spreading exponent fitted from the nanofluid dynamic wetting data agrees with the prediction of the classical hydrodynamics model derived from the bulk viscous dissipation approach. The present study proves that the spreading of the nanofluid droplets is dominated by the bulk dissipation rather than by the local dissipation at the moving contact line.

2.1 Introduction

The attractive and tunable wetting behaviors extend the applications of nanofluids into many scientific and engineering areas; however, the mechanisms of nanofluid dynamic wetting are not well understood [1–6]. Many parameters may be controlled in nanofluids, such as the nanoparticle material, size, shape, and loading, as well as base fluid material. By changing nanoparticles or base fluid, nanofluids exhibit enhanced [3] or reduced [7] thermal conductivity, increased [8, 9] or decreased [10] surface tension, as well as shear-thinning [11] or shear-thickening [12] rheological properties. It can be expected that when nanofluid droplets spread on a solid surface, they will also show different dynamic wetting behaviors. The expectation has been confirmed by the recent reports [13–18]. Wang et al. [13, 14] compared experimentally the spreading behaviors of pure poly (propylene glycol, PPG) and PPG+10 nm silica nanofluids. Their results showed that the wetting radius versus time relation ($R - t$) and the dynamic contact angle versus contact line

© Springer-Verlag Berlin Heidelberg 2016
G. Lu, *Dynamic Wetting by Nanofluids*,
Springer Theses, DOI 10.1007/978-3-662-48765-5_2

velocity relation ($\theta_D - U$) for the pure PPG followed the Newtonian spreading laws; however, the $R - t$ and $\theta_D - U$ relations for the PPG+ silica nanofluids significantly deviated from the Newtonian spreading laws. The deviations were attributed to the fact that adding silica nanoparticles into the solvent led to a shear-thickening rheology. However, a distinct dynamic wetting behavior for nanofluids was reported by Wasan et al. [15–18]. They presented that 8 nm micellar solution and 20-nm silica suspension significantly enhanced the spreading rate as compared with their base fluids. A solid-like ordering structure of nanoparticles was observed near the contact line region using interferometry [15]. Thus, the super-spreading was explained by the structural disjoining pressure due to the self-assembly of nanoparticles in the vicinity of the contact line [15–18]. The super-spreading behavior of nanofluids has been widely used to explain the enhanced drop-wise evaporation [19, 20] and the elevated critical heat flux with nanofluids [21–24]. Another explanation for the super-spreading by nanofluids was that nanoparticles were assumed to settle at the bottom of the droplet, thus reducing the solid–liquid friction and hence facilitating the fluid spreading [25].

The previous studies [13–18] dealt with dynamic wetting behaviors of nanofluids with high nanoparticles fraction. However, high fraction nanofluids are unstable due to the nanoparticle sedimentation, which has become a serious challenge for potential applications of nanofluids. Thus, it is indeed necessary to investigate dynamic wetting characteristics of dilute nanofluids. Recently, Liang et al. [26] proposed that the timescale for the nanoparticles diffusing from the bulk droplet to the contact line region is far larger than that for the droplet spreading, so that the nanoparticle fraction in the contact line region is actually lower than the nominal fraction of bulk droplet. As a result, the shear-thickening nanofluids behave like quasi-Newtonian spreading characteristics. The limited diffusion rate of the nanoparticles was also confirmed for the spreading of gold–water nanofluids on a gold surface via molecular dynamic simulations [27]. For the dilute nanofluids, nanoparticles are harder to diffuse to the contact line region; thus, the super-spreading and shear-thickening spreading may not occur. Unfortunately, the dynamic wetting for the dilute nanofluids has not been reported up to now.

The relations of $\theta_D - U$ and $R - t$ are usually used to describe the dynamic wetting of fluids on solid surfaces. These two relations not only present the wettability of the fluids on the solid surfaces, but also show the energy dissipation mechanisms during the dynamic wetting process [28–31]. Unfortunately, few studies focus on measurements of $\theta_D - U$ and $R - t$ for nanofluids. Without $\theta_D - U$ and $R - t$ experimental data, the dynamic behaviors cannot be accurately described.

This chapter investigates dynamic wetting behaviors of dilute nanofluids by measuring $\theta_D - U$ and $R - t$ data using the droplet spreading method. The effects of nanoparticle material, diameter, and loading, as well as base fluid and substrate material, were examined. The purpose of this work is to reach the following two targets. The first is to answer whether the super-spreading and/or shear-thickening spreading behaviors will also be observed for dilute nanofluids. The second is to address how nanoparticles affect dynamic wetting behaviors in dilute nanofluids, if the two behaviors do not occur.

2.2   Experimental Methods

2.2.1   Experimental Setup

The θ_D–U and R–t data were first measured using a drop shape analyzer (EasyDrop FM40, Krüss GmbH, Hamburg, Germany), as shown in Fig. 2.1a. The spreading process was recorded by a high-speed CCD camera at 60 frames per second. The contact angle and the spreading radius were measured by fitting the droplet profile with the equation $y = a + bx + cx^{0.5} + d/\ln x + e/x^2$ for each picture as shown in Fig. 2.1b. The standard error in the contact angle measurement was $\pm 1°$. The contact line velocity was calculated using $U = df(t)/dt$, where $f(t)$ was fit from the R–t curves.

The surface tension and the θ_D–U curves were then measured using Krüss K100 MK2 (Krüss GmbH, Hamburg, Germany) based on the Wilhelmy plate method, as shown in Fig. 2.2. The substrate was inserted into the liquid reservoir at various velocities. As shown in Fig. 2.2b, the forces imposed on the plate are as follows:

$$F = L \cdot \sigma \cdot \cos\theta_D - \rho gSh, \tag{2.1}$$

where L is the wetting perimeter, ρ is the liquid density, S is the cross-sectional area of the plate, h is the inserted distance, and γ is the liquid–vapor surface tension. The measured F–h curves had good linearity as shown in Fig. 2.2b with the contact angle and then the contact angle is calculated as follows:

$$\theta_D = \arccos\left(\frac{F_0}{L \cdot \sigma}\right), \tag{2.2}$$

(a)   **(b)**

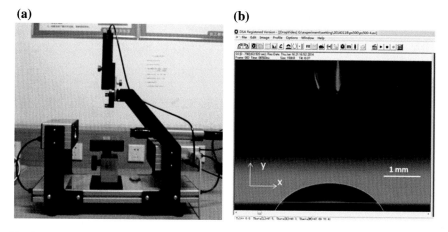

Fig. 2.1 Droplet spreading method: **a** Krüss EasyDrop; **b** Contact angle and spreading radii based on image analysis

(a)   **(b)**

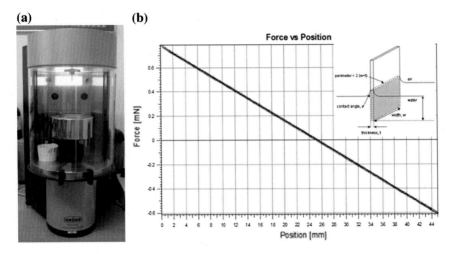

Fig. 2.2 Wilhelmy plate method: **a** Krüss K100 MK2; **b** schematic

Table 2.1 Experimental capacities

Measurement	Method	Company	Type	Capability
$R - t$, $\theta_D - U$	Droplet spreading	Krüss	Krüss FM40 EasyDrop Standard	Resolution: 0.1°; Highest capture spread: 60 frames/s
σ, $\theta_D - U$	Wilhelmy plate	Krüss	Krüss K100 MK2	Max insert: 110 mm; resolution: 0.1 μm; Velocity range: 0.09–500 mm min^{-1}

where F_0 is the intercept in Fig. 2.2b. The surface tension was also measured by a Krüss K100 MK2 using a platinum plate.

The accuracies of the experimental setups are listed in Table 2.1.

2.2.2 Nanofluid Preparation

The measurements of the nanofluid dynamic wetting characteristics required stable nanofluids and clean substrates and beakers. Hence, all the substrates (glass slides, mica slides, and silicon wafers) and the beakers were cleaned with ethanol solution with more than 30-min ultrasonic cleaning. The ultrasonic cleaning procedure was then repeated with acetone and then with the deionized water.

As shown in Table 2.2, the study considered the effects of various nanofluid parameters on the dynamic wetting, including the nanoparticle loading (SiO_2/PDMS500 (polydimethylsiloxane, viscosity of 100 mPa s), $d = 20$ nm, $\varphi = 0$, 0.5, 1, and 2 %), nanoparticle material (SiO_2, TiO_2, Al_2O_3/PDMS500, $d = 20$ nm,

Table 2.2 Nanofluid parameters used for the dynamic wetting tests

Effects	Materials[a]	Diameter (nm)	Base fluids[a]	Loadings (%)	Substrates
Loadings	SiO$_2$	20	PDMS500	0.5	Glass
	SiO$_2$	20	PDMS500	1	Glass
	SiO$_2$	20	PDMS500	2	Glass
Diameters	SiO$_2$	10	PDMS500	1	Glass
	SiO$_2$	15	PDMS500	1	Glass
	SiO$_2$	20	PDMS500	1	Glass
Materials	SiO$_2$	20	PDMS500	1	Glass
	TiO$_2$	20	PDMS500	1	Glass
	Al$_2$O$_3$	20	PDMS500	1	Glass
Base fluids	SiO$_2$	10	PDMS100	1	Glass
	SiO$_2$	10	PDMS500	1	Glass
	SiO$_2$	10	PDMS1000	1	Glass
	SiO$_2$	10	PEG200	1	Glass
	SiO$_2$	10	PEG400	1	Glass
Substrates	SiO$_2$	20	PDMS500	1	Glass[b]
	SiO$_2$	20	PDMS500	1	Mica[c]
	SiO$_2$	20	PDMS500	1	Silicon wafer[d]

[a]Sigma-Aldrich Co., [b]Fisher Co., [c]SPI Supplies, and [d]UMCO

Table 2.3 Nanoparticle properties

Properties	SiO$_2$	SiO$_2$	SiO$_2$	TiO$_2$	Al$_2$O$_3$
Wettability	Hydrophilic				
BET (m^2/g)	90 ± 15	130 ± 25	150 ± 15	50 ± 15	100 ± 15
Diameter (nm)	10 ± 1	15 ± 1	20 ± 1	20 ± 1	20 ± 1
Density (g/l)	50–120	50–120	50–120	–100	–80

$\varphi = 1$ %), nanoparticle diameter (SiO$_2$/PDMS500, $\varphi = 1$ %, $d = 10, 15$, and 20 nm), and base fluid (SiO$_2$/PDMS100, PDMS500, PDMS1000, PEG200 (polyethylene glycol, molecular weight of 200 g/mol), PEG4000, $d = 20$ nm, $\varphi = 1$ %). All the nanoparticles (SiO$_2$, TiO$_2$, and Al$_2$O$_3$) and base fluids (PDMS100, PDMS500, PDMS1000, PEG200, and PEG4000) were Sigma-Aldrich products.

Stable nanofluids were obtained by mixing all the nanofluid suspensions in the ultrasonic cleaner for more than 12 h. The nanofluids were observed to be stable for 48 h. All the measurements were conducted within 1 or 2 h after the nanofluids were prepared. In addition, the dynamic wetting attests to a small droplet lasts only several minutes, which is much shorter than the nanofluid stability time. Hence, the nanofluid stability had no effect on the dynamic wetting experiments. All the tested nanofluid parameters and substrates are listed in Table 2.3.

2.2.3   Experiment Repeatability and Reliability

The repeatability and reliability of experimental method are illustrated by the details
shown in Figs. 2.3, 2.4, and 2.5. Figure 2.3 shows the spreading radius (Fig. 2.3a)
and the dynamic contact angle (Fig. 2.3b) evolution for three tests with the same
PDMS500 liquid within 1 h after the fluid was prepared. Both the spreading radius
and the dynamic contact angle evolution curves almost overlap, indicating the
repeatability and reliability of the experimental method. Figure 2.4 shows seven

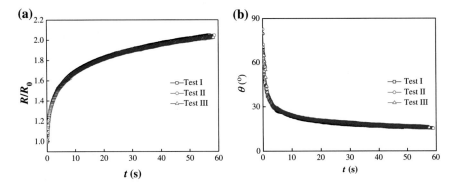

Fig. 2.3 Experiment repeatability testing: three measurements of **a** spreading radius versus time
and **b** dynamic contact angle evolutions for the same PDMS500 liquid within 1 h

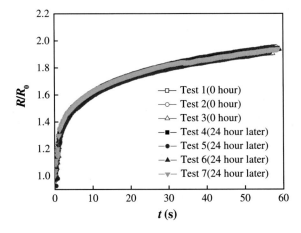

Fig. 2.4 Seven measurements of spreading radius evolution for the same SiO_2/PDMS500
nanofluids ($d = 20$ nm, $\varphi = 0.5$ %): Test Nos. 1–3 were measured immediately when the nanofluids
were well prepared; Test Nos. 4–7 were measured after 24 h when the nanofluids were prepared

Fig. 2.5 Dynamic wetting measurements of Al$_2$O$_3$/PDMS500 nanofluids ($d = 20$ nm, $\varphi = 0.5$ %) which were prepared at different times

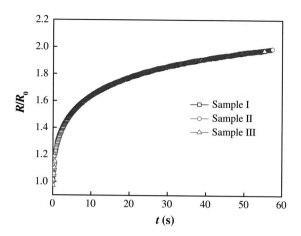

measurements for the same SiO$_2$/PDMS500 nanofluid ($d = 20$ nm, $\varphi = 0.5$ %), in which Test Nos. 1–3 were measured immediately when the nanofluids were prepared, while Test Nos. 4–7 were measured 24 h after the nanofluids were prepared. The consistent results show the good stability of the nanofluids. Figure 2.5 shows the dynamic wetting characteristics of Al$_2$O$_3$/PDMS500 nanofluids ($d = 20$ nm, $\varphi = 0.5$ %) prepared at different times. The overlapping curves show the repeatability of the nanofluid preparation method.

2.3   Results and Discussions

2.3.1   Spreading Behavior of Dilute Nanofluids

Figure 2.6 shows the evolution of the spreading radius for SiO$_2$/PDMS500 nanofluid droplets with various nanoparticle loadings ($\varphi = 0$, 0.5, 1, and 2 %) on the glass slide. Here, R_0 denotes the droplet radius prior to spreading. Since glass slides, mica slides, and silicon wafers are high-energy surfaces, the nanofluid droplets studied in this work were found to spread completely to a thin film on these surfaces. Consequently, the initial radius, R_0, was used for the non-dimensional analysis. Compared with the base fluid ($\varphi = 0$), adding nanoparticles inhibits rather than facilitates the dynamic wetting for the three nanoparticle loadings ($\varphi = 0.5$, 1, and 2 %). The spreading rate and the spreading area decrease with increasing nanoparticle loadings, for example, $R/R_0 = 2.10$ for PDMS500 ($\varphi = 0$) at $t = 50$ s, while $R/R_0 = 1.99$ for $\varphi = 0.5$ %, $R/R_0 = 1.93$ for $\varphi = 1$ %, and $R/R_0 = 1.89$ for $\varphi = 2$ %. The reduced spreading speed and the equilibrium wetting radius were also reported for the impinging of a liquid drop with micron-sized particles on surfaces [32]. For the micron-sized particles, the sedimentation of particles takes place more easily. Thus, an annular particle distribution was observed for higher impinging

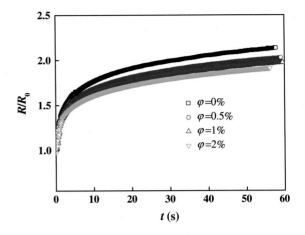

Fig. 2.6 Effects of nanoparticle loadings on the dynamic wetting of nanofluids (SiO$_2$, PDMS500, $\varphi = 0$, 0.5, 1, and 2 %, $d = 20$ nm)

velocities, and the periphery of the drop was always depleted of particles owing to interfacial forces acting on the particles [32]. However, these phenomena do not occur for the spreading of dilute nanofluids.

2.3.2 Individual Parameter Analysis of Nanofluid Dynamic Wetting

To perform the individual parameter analysis, the nanofluids were divided into two groups. One group has different viscosities, while the surface tensions are the same (base fluids: PDMS100, PDMS500, and PDMS1000, nanoparticles: SiO$_2$ with $\varphi = 1$ % and $d = 10$ nm); the other has different surface tensions, while the viscosities are almost the same (base fluids: PDMS100, PEG200, and PEG400, nanoparticles: SiO$_2$ with $\varphi = 1$ % and $d = 10$ nm). Therefore, the effects of viscosity and surface tension can be examined individually. The base fluid properties are shown in Table 2.4.

Table 2.4 Base fluid properties at 20 °C

Base fluids	Density (g/mL)	Surface tension (10^{-3} N/m)	Viscosity (10^{-3} Pa s)
PDMS100[a]	1.06	20.0	100
PDMS500	0.97	20.0	500
PDMS1000	1.09	20.0	1000
PEG200[b]	1.12	37.2	108
PEG400	1.12	43.5	120

[a]PDMS100 is polydimethylsiloxane with viscosity of 100 mPa s
[b]PEG200 is polyethylene glycol with molecular weight of 200 g mol^{-1}

Fig. 2.7 Effects of base fluid
materials on nanofluids
dynamic wetting (SiO$_2$,
$\varphi = 1$ %, $d = 10$ nm)

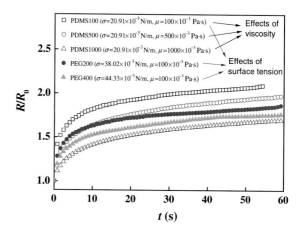

Fig. 2.8 The surface tensions
of nanofluids (SiO$_2$,
$d = 10$ nm, $\varphi = 1$ %)

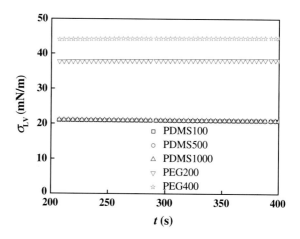

Figure 2.7 shows $(R/R_0) - t$ curves of the five nanofluids. It is found that the
viscosity and surface tension strongly affect the nanofluid dynamic wetting. Both
the spreading velocity and the spreading area decrease with increasing viscosity for
the three PDMS-based nanofluids. For example, R/R_0 is 2.07 for the
PDMS100-nanofluid at $t = 50$ s, while $R/R_0 = 1.94$ for the PDMS500-nanofluid, and
$R/R_0 = 1.65$ for the PDMS1000-nanofluid. For the nanofluids with the same viscosity (PDMS100-based, PEG200-based, and PEG400-based), the spreading
deteriorates with increasing base fluid surface tensions. As shown in Fig. 2.8, the
surface tensions of the five nanofluids measured by the Krüss K100 MK2 remain
unchanged for a long time (longer than the dynamic spreading process). The surface
tension is 20.91 ± 0.04 mN m^{-1} for the three PDMS-based nanofluids,
38.02 ± 0.05 mN m^{-1} for SiO$_2$/PEG200, and 44.33 ± 0.05 mN m^{-1} for SiO$_2$/
PEG400. The effects of surface tension on the dynamic wetting are explained
as follows. The driving force acting on the contact line can be expressed as

Fig. 2.9 Effects of three
nanoparticle diameters on the
dynamic wetting (SiO₂/
PDMS500, $\varphi = 1$ %); inserted
schematic: non-dimensional
radius versus
non-dimensional time

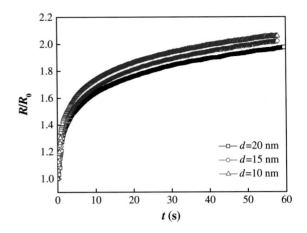

$F = \sigma_{SV} - \sigma_{SL} - \sigma_{LV} \cos \theta_D$, where σ_{SV} and σ_{SL} denote the solid–vapor and solid–liquid interfacial tensions, and σ_{LV} is the liquid–vapor interfacial tension (also referred to as the liquid surface tension). Restated that the nanofluids studied can completely spread on the glass slides, mica slides, and silicon wafers, the dynamic contact angle, θ_D, is always smaller 90° in all spreading experiments. Thus, the increase in σ_{LV} reduces the driving force, which leads to the slower spreading for nanofluids with larger surface tension.

Figure 2.9 shows the effects of the nanoparticle diameter (SiO₂ with $d = 10, 15,$ and 20 nm, the standard deviation of diameter for each nanoparticle is about ±1 nm) on the nanofluid dynamic wetting. The base fluid is PDMS500, and the nanoparticle loading is 1 %. The spreading velocity and spreading area both decrease with increasing nanoparticle diameter. The effects of nanoparticle diameter on the viscosity have been studied experimentally and theoretically [33–36]. These studies demonstrated that increasing the nanoparticle diameter increases the nanofluid viscosity. However, there are no direct evidences to relate the nanoparticle diameter to the surface tension of nanofluids [37–39]. It was reported that the wettability of nanoparticles was responsible for the modified surface tension of nanofluids; adding hydrophilic nanoparticles increases the surface tension of nanofluids, while adding hydrophobic nanoparticles reduces the surface tension [40]. We also measured the surface tensions of nanofluids with the three different nanoparticle diameters. Because the SiO₂ nanoparticles used are hydrophilic, the surface tensions of the three nanofluids are all higher than that of the base fluid, which agrees with the report in Ref. [40]. The results also show that three nanofluids have the same surface tension, indicating that the nanoparticle diameters do not affect the surface tension of nanofluids. Therefore, Fig. 2.9 again confirms that increasing nanofluid viscosity slows down the spreading of nanofluids.

Based on the individual parameter analysis, it is concluded that the super-spreading behavior does not take place, and the viscosity and surface tension are two dominant parameters for the dynamic spreading of dilute nanofluids.

2.3.3   Coupling Effect of Viscosity and Surface Tension

The non-dimensional spreading radius (R/R_0) as a function of the coupling parameter, $t/\mu\sigma_{LV}^2 R_0$, is shown in Fig. 2.10. The use of the coupling parameter is to eliminate the effects of both the surface tension and the viscosity. By eliminating the differences of the surface tension and the viscosity, the original experimental data of R/R_0 versus t (Fig. 2.7) measured for various nanofluids gather together nearly into a single curve. This result indicates that apart from the viscosity and surface tension, there is no other parameter affecting the dynamic wetting of dilute nanofluids. Therefore, for dilute nanofluids, the role of nanoparticles in the dynamic wetting is realized only through modifying the viscosity and surface tension of nanofluids.

The effects of surface tension and viscosity on the dynamic wetting of dilute nanofluids were further tested by adding three different nanoparticle materials (SiO_2, Al_2O_3, and TiO_2) into the same base fluid (PDMS500). The average nanoparticle diameters are 20 nm, and the loadings are 1 % for the three nanoparticle materials. In addition, the three nanoparticles are spherical without any surface treatments. Experimental tests show that three nanofluids have almost the same surface tension and viscosity; thus, R/R_0 versus t curves are expected to coincide with each other. This is verified by the spreading experiments shown in Fig. 2.11.

2.3.4   Mechanisms of Dynamic Wetting in Dilute Nanofluids

2.3.4.1   Spreading Law of Dilute Nanofluids

The R/R_0–t data in Fig. 2.6 were replotted in the dual-logarithmic coordinates, as shown in Fig. 2.12. It is found that the new curves for four nanoparticle loadings are all linear ($R^2 = 0.999$). The curves were then fitted by R–At^α with the results listed

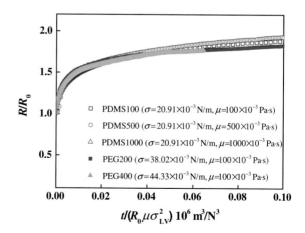

Fig. 2.10 Effects of surface tension and viscosity on the dynamic wetting of nanofluids using non-dimensional analyzing

PDMS100 ($\sigma=20.91\times10^{-3}$ N/m, $\mu=100\times10^{-3}$ Pa·s)

PDMS500 ($\sigma=20.91\times10^{-3}$ N/m, $\mu=500\times10^{-3}$ Pa·s)

PDMS1000 ($\sigma=20.91\times10^{-3}$ N/m, $\mu=1000\times10^{-3}$ Pa·s)

PEG200 ($\sigma=38.02\times10^{-3}$ N/m, $\mu=100\times10^{-3}$ Pa·s)

PEG400 ($\sigma=44.33\times10^{-3}$ N/m, $\mu=100\times10^{-3}$ Pa·s)

Fig. 2.11 Effects of three
nanoparticle materials on the
dynamic wetting (PDMS500,
$\varphi = 1$ %, $d = 20$ nm)

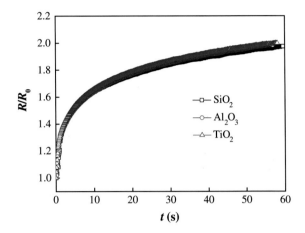

Fig. 2.12 Linear fitting in the
logarithmic coordinates for
nanofluid dynamic wetting
with various nanoparticle
loadings (SiO$_2$, PDMS500,
$\varphi = 0$, 0.5, 1, and 2 %,
$d = 20$ nm)

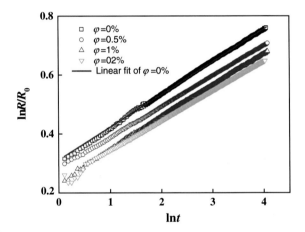

in Table 2.5. The spreading exponent, α, can be used to determine the energy dissipation mechanism. If the dynamic wetting is dominated only by the energy dissipation occurred in the vicinity of contact line, referred to as local dissipation, the molecular kinetic theory (MKT) predicts $\alpha = 1/7$ [31]. On the contrary, the hydrodynamics model [28–30] assumed that the viscous dissipation in the bulk droplet dominates the dynamic wetting, which predicts $\alpha = 1/8$ in the gravitational spreading regime and $\alpha = 1/10$ in the capillary spreading regime.

The fitted spreading exponents, α, are close to 1/10 ($\Delta = 11$ % for $\varphi = 0$ %, $\Delta = 4$ % for $\varphi = 0.5$ %, $\Delta = 6$ % for $\varphi = 1$ %, and $\Delta = 0$ % for $\varphi = 0$ %), which meets the prediction of the classical hydrodynamics model derived from the bulk viscous dissipation approach for Newtonian flows. Therefore, the bulk dissipation dominates the dynamic wetting of SiO$_2$/PDMS500 nanofluids. It should be noted that the super-spreading of nanofluids is controlled by the local energy dissipation, because the super-spreading comes from the structural disjoining pressure due to the

Table 2.5 Spreading laws of SiO$_2$-PDMS500 with various loadings

φ (%)	A	α	R^2
0	1.366	0.111	0.999
0.5	1.329	0.104	0.999
1	1.277	0.106	0.999
2	1.273	0.100	0.998

nanoparticle self-assembly near the contact line region. Thus, the bulk dissipation mechanism for dynamic wetting of dilute nanofluids indicates that the nanoparticles do have no enough time to diffuse to the contact line region during the dynamic wetting so that the self-assembly of nanoparticles cannot take place in the present dilute nanofluids. According to the hydrodynamics model, the spreading exponents of about 1/10 also indicate that the capillary force is the only driving force, while the viscous force is the only resistance force. Thus, the role of adding nanoparticles in dilute nanofluids is to change the viscosity and surface tension of nanofluids, and then, these two physical properties affect the dynamic wetting.

2.3.4.2   Nanoparticle Behaviors During Dilute Nanofluid Dynamic Wetting

Figure 2.13 shows the dynamic wetting of SiO$_2$/PDMS500 nanofluids (d = 20 nm, φ = 1 %) on glass, silicon, and mica slides. The three substrates are chemically and physically homogeneous. The AFM scanning results show that all three surfaces have nanoscale roughness (the root-mean-square roughness is 0.231 nm for glass slide surface, 0.125 nm for the silicon, and 0.137 nm for the mica), as shown in Fig. 2.14. Therefore, the substrates are smooth and ideal surfaces for dynamic wetting [41], so that the dynamic wetting is only affected by the solid surface energy of these three substrates.

Fig. 2.13 Contact line velocity versus the contact angle for three substrates

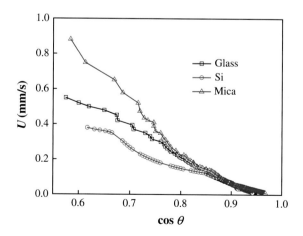

(a)   **(b)**   **(c)**

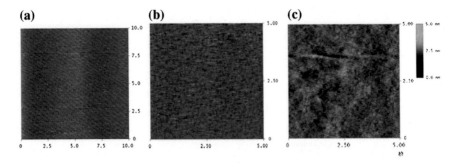

Fig. 2.14 Surface roughness scanning of three substrates using AFM: **a** glass slide; **b** sillicon wafor; **c** mica slide

According to MKT [31], the ratio of the contact line velocity to the cosine of contact angle is proportional to the ratio of the liquid–vapor surface tension and the solid–liquid friction coefficient,

$$U/\cos\theta_D \sim \sigma_{LV}/\zeta, \tag{2.3}$$

where ζ is the solid–liquid "friction" coefficient in the MKT model, characterizing the intermolecular interactions between the solid substrate and liquid phase. The $U-\cos\theta_D$ curves for the three substrates (Fig. 2.10) are linear over most of the droplet spreading time. The ratios of σ_{LV}/ζ can be obtained from the slopes of the $U-\cos\theta_D$ curves. With $\sigma_{LV} = 20.91$ mN m^{-1} for SiO$_2$/PDMS500 nanofluids, ζ is 14.95 Pa s for the glass slides, $\zeta = 19.13$ Pa s for the silicon wafer slides, and $\zeta = 9.03$ Pa s for the mica slides. The "friction" coefficient ζ is used to demonstrate that there are no nanoparticles deposited in the vicinity of contact line region during the dynamic wetting. If the self-assembly of nanoparticles occurs in the vicinity of contact line region, the contact line will move on the "SiO$_2$ solid surfaces," no matter what the substrate is; thus, the value of ζ for glass, silicon, and mica slides should equal with each other. The different ζ indicates that the self-assembly of nanoparticles does not occurs. The results provide an indirect evidence to confirm that the dilute nanofluid dynamic wetting is dominated by the bulk dissipation.

2.3.4.3 Newtonian Dynamic Wetting Behaviors of Dilute Nanofluids

It is restated that the 1/10 spreading exponent is derived from the hydrodynamics model for the Newtonian dynamic wetting. Thus, the dynamic wetting of dilute nanofluids behaves like that of Newtonian fluids, which is further confirmed by measuring the rheological properties of nanofluids with four nanoparticle loadings (PDMS500, SiO$_2$ with $\varphi = 0$, 0.5, 1, and 2 %, $d = 20$ nm), as shown in Fig. 2.15. There is still a debate about whether nanofluids exhibit Newtonian or non-Newtonian behavior [42–48]. Chen et al. [49] and Yu et al. [50] stated that the

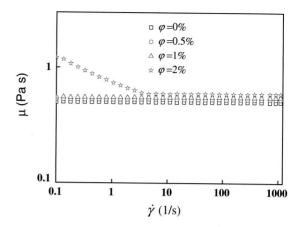

Fig. 2.15 Relation of dynamic viscosity versus shear rate for nanofluids with four nanoparticle loadings (SiO$_2$, PDMS500, φ = 0, 0.5, 1, and 2 %, d = 20 nm)

Newtonian or non-Newtonian rheology of nanofluids depends strongly on the nanoparticle volume fraction. The volume-fraction-dependent rheology of nanofluids was explained on the molecular level by considering the effects of the nanoparticle motion and aggregation [40]. In Wang et al.'s work [13, 14], the mass fraction of silica nanoparticles in PPG was 7.5 and 10 %, far higher than the present loadings, which leads to the shear-thickening rheology and hence to the shear-thickening dynamic wetting behavior. However, for the present dilute nanofluids, the nanoparticles distribute uniformly in the bulk liquid. This homogeneous nanoparticle distribution reduces the nanoparticle aggregation in the bulk liquid, leading to the Newtonian-like dynamic wetting behavior.

2.4 Conclusions

The variations of wetting radius and contact angle with time were measured using the droplet spreading method to study the dynamic wetting behaviors of dilute nanofluids. Various effects, such as the nanoparticle material, loading, and diameter, the base fluid, and the substrate, were considered in this study. The main conclusions are as follows.

1. The nanoparticles inhibit rather than facilitate the dynamic wetting of dilute nanofluids. Both the contact line velocity and the spreading area decrease with increasing loading. The individual parameter analysis shows that the deterioration in dynamic wetting for dilute nanofluids can be attributed to the increase in either surface tension or viscosity due to adding nanoparticles into the base fluid.

2. The spreading exponent fitted from the nanofluid dynamic wetting data is found to be very close to 0.1, which meets the prediction of the classical hydrodynamics model derived from bulk viscous dissipation approach for Newtonian

flows. This is because the nanoparticles in dilute nanofluids are uniformly distributed in the bulk liquid. The homogeneous nanoparticle distribution reduces the nanoparticle aggregation in the bulk liquid and the self-assembly in the contact line region, leading to a Newtonian-like dynamic wetting behavior.

3. It is interesting that once the effects of the surfaces tension and viscosity are both eliminated using the non-dimensional analysis, the wetting radius versus spreading time (R–t) curves for all the nanofluid droplets overlap with each other. This result and the Newtonian-like behavior demonstrate that the dynamic wetting of dilute nanofluids is dominated by the bulk dissipation.

The present results provide a better understanding and direct evidences of the bulk dissipation mechanism for the dilute nanofluid dynamic wetting. The finding of Newtonian-like behavior in dilute nanofluids also provides a guideline for building theoretical models of the dynamic wetting of dilute nanofluids.

References

1. Chakraborty S, Padhy S (2008) Anomalous electrical conductivity of nanoscale colloidal suspensions. ACS Nano 2:2029–2036
2. Trisaksri V, Wongwises S (2007) Critical review of heat transfer characteristics of nanofluids. Renew Sust Energ Rev 11:512–523
3. Branson BT, Beauchamp PS, Beam JC et al (2013) Nanodiamond nanofluids for enhanced thermal conductivity. ACS Nano 7:3183–3189
4. Choi SUS (1995) Enhancing thermal conductivity of fluids with nanoparticles. In: Developments and application of non-newtonian flows. ASME, New York
5. Choi SUS (2009) Nanofluids: from vision to reality through research. J Heat Transfer 131:033106–033111
6. Cheng LS, Cao DP (2011) Designing a thermo-switchable channel for nanofluidic controllable transportation. ACS Nano 5:1102–1108
7. Michaelides EE (2013) Transport properties of nanofluids. A critical review. J Non-Equilib Thermodyn 38:1–79
8. Moosavi M, Goharshadi EK, Youssefi A (2010) Fabrication, characterization, and measurement of some physicochemical properties of ZnO nanofluids. Int J Heat Mass Transfer 31:599–605
9. Tanvir S, Li Q (2012) Surface tension of nanofluid-type fuels containing suspended nanomaterials. Nanoscale Res Lett 7:226–236
10. Vafaei S, Purkayastha A, Jain A (2009) The effect of nanoparticles on the liquid-gas surface tension of Bi_2Te_3 nanofluids. Nanotechnology 20:185702
11. Chen HS, Ding YL, Lapkin A (2009) Rheological behaviour of nanofluids containing tube rod-like nanoparticles. Powder Tech 194:132–141
12. Lee YS, Wagner NJ (2003) Dynamic properties of shear thickening colloidal suspensions. Rheol Acta 42:199–208
13. Wang XD, Lee DJ, Peng XF et al (2007) Spreading dynamics and dynamic contact angle of non-Newtonian fluids. Langmuir 23:8042–8047
14. Wang XD, Zhang Y, Lee DJ (2007) Spreading of completely wetting or partially wetting power-law fluid on solid surface. Langmuir 23:9258–9262
15. Wasan DT, Nikolov AD (2003) Spreading of nanofluids on solids. Nature 423:156–159

16. Kondiparty K, Nikolov AD, Wu S (2011) Wetting and spreading of nanofluids on solid surfaces driven by the structural disjoining pressure: statics analysis and experiments. Langmuir 27:3324–3335
17. Kondiparty K, Nikolov AD, Wasan DT et al (2012) Dynamic spreading of nanofluids on solids. Part I: experimental. Langmuir 28:14618–14623
18. Liu KL, Kondiparty K, Nikolov AD et al (2012) Dynamic spreading of nanofluids on solids. Part II: modeling. Langmuir 28:16274–16284
19. Sefiane K, Bennacer R (2009) Nanofluids droplets evaporation kinetics and wetting dynamics on rough heated substrates. Adv Colloid Interface Sci 147–148:263–271
20. Moffat JR, Sefiane K, Shanahan MER (2009) Effect of TiO_2 nanoparticles on contact line stick-slip behavior of volatile drops. J Phys Chem B 113:8860–8866
21. Murshed SMS, Nieto de Castro CA, Lourenco MJV et al (2007) A review of boiling and convective heat transfer with nanofluids. Renew Sust Energy Rev 15:2342–2354
22. Wen DS (2008) Mechanisms of thermal nanofluids on enhanced critical heat flux (CHF). Int J Heat Mass Transfer 51:4958–4965
23. Kim SJ, Bang IC, Buongiorno J et al (2006) Effects of nanoparticle deposition on surface wettability influencing boiling heat transfer in nanofluids. Appl Phys Lett 89:153107
24. Sefiane K (2006) On the role of structural disjoining pressure and contact line pinning in critical heat flux enhancement during boiling of nanofluids. Appl Phys Lett 89:044106
25. Sefiane K, Skilling J, MacGillivary J (2008) Contact line motion and dynamic wetting of nanofluid solutions. Adv Colloid Interface Sci 138:101–120
26. Liang ZP, Wang XD, Lee DJ et al (2009) Spreading dynamics of power-law fluid droplets. J Phys Condens Matter 21:464117
27. Lu G, Hu H, Duan YY et al (2013) Wetting kinetics of water nano-droplet containing non-surfactant nanoparticles: a molecular dynamics study. Appl Phys Lett 103:253104
28. Huh C, Mason SG (1977) Steady movement of a liquid meniscus in a capillary tube. J Fluid Mech 81:401–419
29. Dussan VEB (1976) The moving contact line: the slip boundary conditions. J Fluid Mech 76:665–684
30. Tanner LH (1979) The spreading of silicone oil drops on horizontal surfaces. J Phys D 12:1473–1484
31. Blake TD, Haynes JM (1969) Kinetics of liquid/liquid displacement. J Colloid Interface Sci 30:421–423
32. Nicolas M (2005) Spreading of a drop of neutrally buoyantsuspension. J Mech Fluid 545:271–280
33. Chandrasekar M, Suresh S, Chandra BA (2010) Experimental investigations and theoretical determination of thermal conductivity and viscosity of Al_2O_3/water nanofluids. Exp Thermal Fluid Sci 34:210–216
34. Nguyen CT, Desgranges F, Galanis N et al (2008) Viscosity data for Al_2O_3-water nanofluid-hysteresis: is heat transfer enhancement using nanofluids reliable. Int J Therm Sci 47:103–111
35. Lee JH, Hwang KS, Janga S et al (2008) Effective viscosities and thermal conductivities of aqueous nanofluids containing low volume concentrations of Al_2O_3 nanoparticles. Int J Heat Mass Transfer 51:2651–2656
36. Prasher R, Song D, Wang J et al (2006) Measurements of nanofluid viscosity and its implications for thermal applications. Appl Phys Lett 89:133108–133111
37. Murshed SM, Tan SH, Nguyen NT (2008) Temperature dependence of interfacial properties and viscosity of nanofluids for droplet-based microfluidics. J Phys D Appl Phys 41:085502
38. Radiom M, Yang C, Chan WK (2010) Characterization of surface tension and contact angle of nanofluids. Proc SPIE 7522:75221D
39. Liu Y, Kai D (2012) Investigations of surface tension of binary nanofluids. Adv Mater Res 347–353:786–790
40. Lu G, Duan YY, Wang XD (2014) Surface tension, viscosity, and rheology of water-based nanofluids: a microscopic interpretation on the molecular level. J Nanopart Res 16:2564

41. Chen P (2005) Molecular interfacial phenomena of polymers and biopolymers. In: Grundke K (ed) Surface-energetic properties of polymers in controlled architecture. Woodhead, Cambridge, pp 323–374
42. Chen HS, Ding YL, He YR et al (2007) Rheological behaviour of ethylene glycol based titania nanofluids. Chem Phys Lett 444:333–337
43. Susan-Resiga D, Socoliuc V, Boros T et al (2012) The influence of particle clustering on the rheological properties of highly concentrated magnetic nanofluids. J Colloid Interface Sci 373:110–115
44. Wang XW, Xu XF, Choi SUS (1999) Thermal conductivity of nanoparticle-fluid mixture. J Thermophys Heat Transfer 13:474–480
45. Chen HS, Ding YL, Lapkin A (2009) Rheological behaviour of nanofluids containing tube rod-like nanoparticles. Powder Technol 194:132–141
46. Ding Y, Alias H, Wen D et al (2006) Heat transfer of aqueous suspensions of carbon nanotubes (CNT nanofluids). Int J Heat Mass Transfer 49:240–250
47. Kole M, Dey TK (2011) Effect of aggregation on the viscosity of copper oxide-gear oil nanofluids. Int J Thermal Sci 50:1741–1747
48. Kim S, Kim C, Lee WH et al (2011) Rheological properties of alumina nanofluids and their implication to the heat transfer enhancement mechanism. J Appl Phys 110:34316
49. Chen HS, Ding YL, Tan CQ (2007) Rheological behaviour of nanofluids. New J Phys 9:367
50. Yu W, Xie H, Chen L et al (2009) Investigation of thermal conductivity and viscosity of ethylene glycol based ZnO nanofluids. Thermo Chimica Acta 491:92–99

Chapter 3
Local Dissipation in Nanofluid Dynamic Wetting: Effects of Structural Disjoining Pressure

Abstract In this chapter, the dynamic wetting of nanofluid droplets is examined using molecular dynamics simulations. The main purpose is to provide the definition and the explanation of local dissipation in nanofluid dynamic wetting due to the nanoparticle self-assembly. In addition, the magnitude of structure disjoining pressure is compared with the unbalanced Young's stress acting at the contact line in the second part to examine the effects of structural disjoining pressure due to the nanoparticle self-assembly in the vicinity of the contact line region. The microscopic mechanism of the contact line motion is investigated on the atomic level.

3.1 Introduction

The addition of nanoparticles makes the wetting kinetics of nanofluids more complicated due to the additional particle–particle, particle–substrate, and particle–fluid interactions. To distinguish different roles of nanoparticles in the nanofluid dynamic wetting, we defined two energy dissipations in this dissertation, bulk dissipation and local dissipation. The roles of these two dissipations on the nanofluid dynamic wetting have been revealed experimentally in Chap. 2. The following chapters will discuss the mechanisms of these two energy dissipations. The bulk dissipation is referred to the effects of nanoparticle random motion in the bulk liquid, while the local dissipation characterizes the self-assembly and sedimentation of nanoparticles in the vicinity of contact line region. The self-assembly of nanoparticles within the bulk liquid, at the solid–liquid and liquid–vapor interfaces, and/or in the vicinity of the contact line region greatly affects the wettability of nanofluids, which was reported by Wasan and Nikolov [1] with a new term of "super-spreading" by nanofluids. The enhancement in the wettability by nanofluids was attributed to the solid-like ordering structure of nanoparticles near the contact line region which was observed in their experiments using interferometry. This solid-like ordering structure stemming from the settlement and assembly of nanoparticles gives rise to a structural disjoining pressure in the vicinity of the contact line, which alters the force balance

© Springer-Verlag Berlin Heidelberg 2016
G. Lu, *Dynamic Wetting by Nanofluids*,
Springer Theses, DOI 10.1007/978-3-662-48765-5_3

near the contact line and enhances the wettability of nanofluids [1–5]. The super-spreading behavior of nanofluids was also reported in aluminum–ethanol nanofluids and SWCNT-water [6–8]. The self-assembly of nanoparticles and the structural disjoining pressure-induced super-spreading behavior of nanofluids were widely used to explain the enhancements in droplet evaporation [9, 10] and critical heat flux of nanofluid boiling [11–14].

Although experimental observations [1, 4] have been reported for the effects of ordering of nanoparticles inside the micron-sized wedge film near a contact line on the static wettability (equilibrium stage) of nanofluid droplets, no direct evidence has been provided for the fast dynamic wetting of nanofluid droplets, a process important to many nanofluidic devices. Molecular dynamic (MD) simulations have been recognized as a powerful tool in examining the transport phenomena of nanoparticles in bulk liquids or at interfaces [15–20]. Drazer et al. [21, 22] and Li and Drazer [23] investigated the mobility of a single nanoparticle transporting in a cylindrical nanochannel filled with nitrogen gas. Kim et al. [24] simulated the ring-like deposition of nanoparticles from a drying droplet using a Monte Carlo scheme. To the best of our knowledge, no MD simulations have been reported for dynamic wetting of nanofluids on solid surfaces. In addition, most MD simulations focused on nanofluids containing Lennard-Jones (LJ) particles in an LJ fluid, a system that is quite different from real nanofluids and is hence impossible to mimic important wettability-related properties (e.g., surface tension, density, and viscosity) of nanofluids.

In this chapter, the dynamic wetting of nanofluid droplets is examined using MD simulations. The main purpose was to provide the definition and the explanation of local dissipation in nanofluid dynamic wetting due to the nanoparticle self-assembly. In addition, the magnitude of structure disjoining pressure was compared with the unbalanced Young's stress acting at the contact line in the second part to examine the effects of structural disjoining pressure due to the nanoparticle self-assembly in the vicinity of the contact line region. The microscopic mechanism of the contact line motion was investigated by MD simulations.

3.2 Simulation Setups

Figure 3.1 shows the simulation procedure of nanofluid dynamic wetting. The water cubes were equilibrated at NVT ensembles (N is the number of atoms, V the volume, and T the temperature) to generate the cylindrical water droplets (diameter $d = 10$ nm, length $l = 1.64$ nm, and 4500 water molecules). The substrates are Au (100) surfaces ($61.35 \times 1.64 \times 1.64$ nm^3 fcc crystal total of 9600 gold atoms). The gold nanoparticles ($0.8 \times 0.8 \times 0.8$ nm^3, 32 gold atoms per particle) were calibrated at 300 K and then absorbed spontaneously by a water droplet until the nanofluid system reached its minimum energy where the nanoparticles were randomly distributed inside the drop. Four gold nanoparticle loadings (number of nanoparticles $n = 0, 9, 18,$ and 27, corresponding to particle volume fraction $\varphi = 0, 3.43, 6.77,$ and

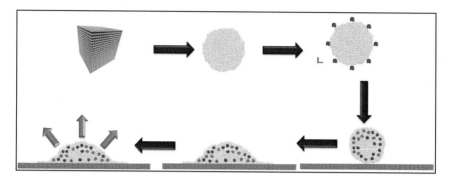

Fig. 3.1 Schematic of simulation procedure for nanofluid droplet spreading

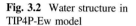
Fig. 3.2 Water structure in TIP4P-Ew model

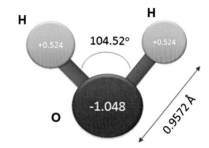

9.81 % in volume fraction) were considered in this study. In present simulation, the particle–water interaction energy ε was changed to mimic various nanoparticle types with different wettabilities (the wettability increases with increasing ε) while keeping ε of substrate–water the same. All the nanoparticles are non-surfactant in present simulations. The newly formed nanofluid droplets then spread on an Au (100) surface until the equilibrium contact angles were achieved. The four-point TIP4P-Ew water model [25] was used to describe the water–water interactions, where the particle–particle particle–mesh (PPPM) technique was used to compute the long-range Coulombic forces and the SHAKE algorithm was applied to keep the water molecules rigid [26]. The model parameters in TIP4P-Ew water model are shown in Fig. 3.2, where the OH bond length is 0.9572 Å, the HOH angle is 104.52°, the charge of O^{2-} is −1.048 e, and the charge of H^+ is 0.524 e. The embedded-atom method (EAM) [27] was applied for the gold–gold interactions, which occur among particle–particle, particle–substrate, and substrate–substrate atoms. A 12-6 LJ potential with $\sigma = 3.1$ Å [28] and a cutoff distance of 9 Å were used to describe the water–gold interactions. The simulation errors were calculated from using four different random velocity seeds (87,287, 36,986, 662,952, and 153,388). The droplet spreading simulations were performed using LAMMPS [29] under *NVT* ensembles at $T = 300$ K. A time step of 1 fs was used in all cases. The spreading radii are defined using free surface profile fitting method. The droplet

Fig. 3.3 Model validation.
a R–t relation fitted from the
spreading radius versus
spreading time. **b** Solid–liquid
friction coefficient in MKT
fitted from the θ–U data.
Reprinted from Ref. [33],
with kind permission from
ASME

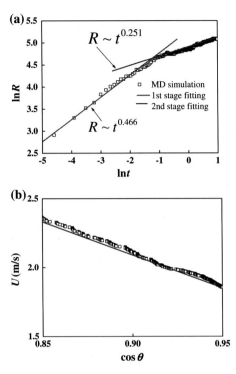

above the substrate was divided into several layers. Within each layer at a given t, the density of drop atoms is calculated as a function of distance from the center of mass of the layer. Density data are integrated until 98 % of drop atoms in the layer are accounted for; the distance at which this occurs is $R(t)$ for the layer. The criterion of less than 100 % is used to avoid erroneous data generated by single water molecules breaking away from the droplet. With various layers of $R(t)$, the dynamic contact angle can be measured by linear fitting of the instantaneous drop radius of the first six layers of water molecules above the gold substrate surface.

The model validation is shown in Fig. 3.3. For the wetting of a liquid droplet on a solid surface, models have been developed to connect wetting kinetics to relevant driving forces and dissipation mechanisms using the power-law form of the drop spreading radius, R, versus time t, i.e., R–t^{α}, where α is determined by the specific dissipation mechanism. Alternatively, expressions connecting the contact line velocity U to the dynamic contact angle θ can be used to identify which mechanisms are dominant in determining the contact line kinetics. Dynamic models for inert wetting include those identifying liquid viscous forces as the primary resistance to spreading, i.e., the hydrodynamic model [30], as compared to those identifying solid–liquid molecular frication at the contact line as the dominant dissipation mechanism, i.e., MKT [31]. The hydrodynamic model gives $\alpha = 1/10$ for a spherical drop and $\alpha = 1/7$ for a cylindrical geometry at the capillary regime and the MKT yields $\alpha = 1/7$ for spherical and $\alpha = 1/5$ for cylindrical, respectively.

In this study, the curve-fitted power α of the cylindrical water drop wetting on a Au (100) surface is 0.251, close to that of $\alpha = 1/5$ from the MKT. This result implies that molecular adsorption and desorption of water to and from the Au(100) surface is very likely to be the dominant dissipation mechanism during wetting. In the MKT [31], the relation of the contact line velocity and dynamic contact angle gives $U/\cos\theta - \sigma_{LV}/\zeta$, where ζ is the solid–liquid friction coefficient. It is noted that the calculated friction coefficient of pure water is 0.0143 Pa s, consistent with the reported value of 0.01 Pa s for water on a polyethylene terephthalate substrate [32].

3.3 Effects of Nanoparticle Motion

Figure 3.4 shows the snapshots of a pure water droplet and three water–gold nanofluid droplets with particle volume fractions of 3.43, 6.77, and 9.81 %, respectively, wetting on a Au(100) surface of 300 K at $t = 10$ ns. The water molecules are set transparent for a better visualization of the gold nanoparticles within the droplets from (b) to (d). It can be observed that the drop spreading radius decreases with the increasing nanoparticle volume fraction. The addition of nanoparticles inhibits rather than enhances the spreading of nanoscale droplet during the nanosecond spreading process. The nanoparticles distribute randomly in the droplet at $t = 10$ ns (equilibrium stage) for nanofluids of all three particle loadings. No sign of solid-like ordering structure is observed in the vicinity of the contact line during nanofluid droplet spreading, different from those reported in macroscopic experiments [9]. To examine the effect of self-assembly of nanoparticles, the diffusion timescale of nanoparticles was estimated by the Einstein diffusion theory [34],

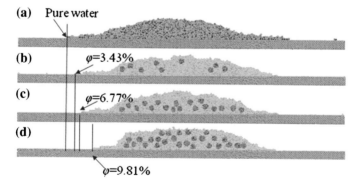

Fig. 3.4 Snapshots of nanoscale water droplet containing Au nanoparticles of various volume fractions spreading on a Au(100) surface of 300 K at $t = 10$ ns. **a** Pure water. **b** Nanofluid of $\varphi = 3.43$ % ($n = 9$). **c** $\varphi = 6.77$ % ($n = 18$). **d** $\varphi = 9.81$ % ($n = 27$). Reprinted with permission from [37]. Copyright 2013, AIP Publishing LLC

$$D = \frac{k_B T}{3\pi\mu d} \tag{3.1}$$

where k_B is the Boltzmann constant 1.38×10^{-23} J/K, T is the temperature, μ is the viscosity of water, and d is the nanoparticle diameter. The diffusion time for one nanoparticle to move from the center of the droplet to the interface is given as follows:

$$t_d = \frac{\pi R^2}{D} \approx 160\,\text{ns} \tag{3.2}$$

at 300 K, much longer than the entire spreading process of approximately 10 ns. This implies that not enough time is available for nanoparticles to self-assemble into ordered structures near the contact line region during the spreading process. Consequently, the spreading enhancement of nanofluids due to the presence of structural disjoining pressure as a result of nanoparticle ordering is not valid mechanism in this study. Limited diffusion rate of nanoparticles in the contact line region as nanofluid spreads was also reported when 10-nm silica nanoparticles suspended in a polyethylene glycol solution [35, 36]. It is noted here that the droplet sizes used in most nanofluid wetting studies were several millimeters in diameter [1, 4], corresponding to a much longer spreading time during which solid-like ordering structures of nanoparticles may occur. In addition, the preview studies mainly focused on the effects of nanoparticles on the nanofluid static wettability, in which the nanoparticle might have enough time to transport to the vicinity of the contact line region.

In a microfluidic system, the dynamic spreading of a droplet is mainly controlled by the viscosity and surface tension of the fluid. To examine the effects of these two properties on dynamic spreading, the dimensionless drop spreading radius, $R' = R/R_0$, as a function of the dimensionless time $t' = \sigma t/\mu_{eff} R_0$ is plotted in Fig. 3.5. Here, R_0 is the initial drop radius prior to spreading, σ is the surface tension of water, and μ_{eff} is the effective viscosity of the nanofluid, calculated using

$$\mu_{eff} = \mu(1 + 2.5\varphi) \tag{3.3}$$

to eliminate the effect of viscosity in droplet spreading kinetics. For fluids of equal surface tension, the R' versus t' curves are expected to collapse into one master curve [38]. However, as shown in Fig. 3.4, both the dynamics and the equilibrium spreading radii decrease with the presence of nanoparticles and further decrease with the increasing nanoparticle volume fraction, indicating a change in surface tension of nanofluids with particle loading.

To evaluate the surface tension of nanofluids, two NVT systems, one with a nanofluid film of two free surfaces and the other with only a bulk nanofluid, were

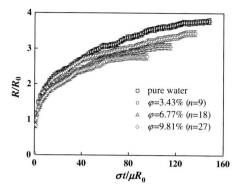

Fig. 3.5 Dimensionless spreading radius as a function of the dimensionless time for cylindrical water nanofluid droplets containing Au nanoparticles of volume fractions of 0, 3.43, 6.77, and 9.81 % on a Au(100) substrate. Reprinted with permission from [37]. Copyright 2013, AIP Publishing LLC

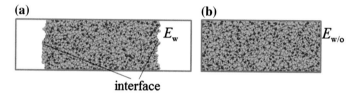

Fig. 3.6 Surface tension calculation based on excess energy method. Reprinted from Ref. [33], with kind permission from ASME

simulated, as shown in Fig. 3.6. The time-averaged excess energy of these two systems, $\langle E_\mathrm{w} \rangle - \langle E_\mathrm{w/o} \rangle$, led to the surface tension,

$$\sigma_\mathrm{LV} = \frac{\langle E_\mathrm{w} \rangle - \langle E_\mathrm{w/o} \rangle}{A_\mathrm{interface}} \tag{3.4}$$

for pure water and nanofluids with different particle loadings. Here, the subscripts w and w/o denote with and without liquid–vapor interfaces, respectively. In MD simulations, surface tension can also be calculated from pressure based on the Young–Laplace equation [39] but is avoided here due to large pressure fluctuations. Table 3.1 summarizes the calculated liquid–vapor surface tensions based on the excess energy method for pure water, and nanofluids of $\varphi = 3.43$, 6.77, and 9.81 %. For pure water, the calculated surface tension is $\sigma = 0.06793 \pm 0.0043$ N/m at $T = 300$ K, within 6 % error compared with the experimental value of $\sigma = 0.072$ N/m [40]. As the nanoparticle loading increases, the surface tension of nanofluids increases from 0.1002 ± 0.0065 N/m for $\varphi = 3.43$ % to 0.1175 ± 0.0075 N/m for $\varphi = 9.81$ %. The increase in surface tension of liquids with non-surfactant nanoparticles was also reported by Tanvir and Li [41] and Moosavi

Table 3.1 Liquid–vapor surface tension of water and water–gold nanofluids of different volume fractions based on the free energy method

	Pure water	$n = 9$	$n = 18$	$n = 27$
φ	0	3.43 %	6.77 %	9.81 %
$\langle E_w \rangle$ (eV)	-1789.081 ± 3.343	-2841.828 ± 5.399	-3885.474 ± 8.548	-4943.323 ± 10.875
$\langle E_{w/o} \rangle$ (eV)	-1796.242 ± 3.481	-2853.841 ± 5.993	-3910.417 ± 7.820	-4969.034 ± 10.435
σ_{LV} (N/m)	0.0679 ± 0.0043	0.1002 ± 0.0065	0.1109 ± 0.0073	0.1175 ± 0.0075

Reprinted with permission from [37]. Copyright 2013, AIP Publishing LLC

et al. [42] as a result of the van der Waals forces between particles at the liquid–gas interface.

The unbalanced Young's stress,

$$F = \sigma_{SV} - \sigma_{SL} - \sigma_{LV} \qquad (3.5)$$

where subscripts S, L, and V denote solid, liquid, and vapor phases, respectively, is the resultant force triggering the outspreading motion of the contact line. As the liquid–vapor surface tension increases with the increasing volume fraction of non-surfactant nanoparticles, the unbalanced Young's stress decreases, corresponding to a slower contact line velocity and a smaller equilibrium spreading radius. The wettability deterioration due to the increase in surface tension was also reported by Moffat et al. [10], Kwok and Neumann [43], and Chen et al. [44].

Figure 3.7 illustrates $U - \cos\theta$ for nanofluid droplets with particle volume fractions of 0, 3.43, 6.77, and 9.81 %. The results show good linearity ($R^2 > 0.989$) of $U - \cos\theta$ for $t > 3$ ns ($\cos\theta > 0.8$ corresponding to 36.9°, consistent with the validity of the MKT for wetting scenarios with a small contact angle). The slope attained from the linear fitting of $U - \cos\theta$ gives $\sigma_{LV}/\zeta = 5.294 \pm 0.258$ for

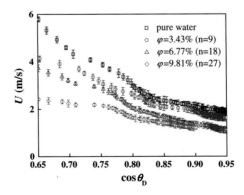

Fig. 3.7 The contact line velocity, U, as a function of $\cos\theta$ (θ the dynamic contact angle) for water–gold nanofluid droplets with particle volume fractions of 0, 3.43, 6.77, and 9.81 %. Reprinted with permission from [37]. Copyright 2013, AIP Publishing LLC

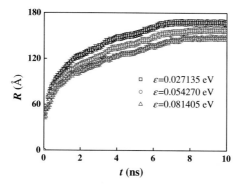

Fig. 3.8 The spreading radius as a function of spreading time for nanofluid droplets with different particle wettabilities for particle volume fraction of $\varphi = 9.81$ %. Reprinted with permission from [37]. Copyright 2013, AIP Publishing LLC

nanofluids of particle loading up to ~ 10 %. Similar σ_{LV}/ζ ratios for pure water and nanofluids of various particle loadings (the standard deviation of 0.258) indicate that the solid–liquid friction coefficient increases with the increasing liquid–vapor surface tension. The MKT assumes that the viscous dissipation in a bulk liquid is negligible while focusing on the local dissipation in the vicinity of the contact line so that the friction coefficient, ζ, is affected only by surface tension. A larger surface tension due to a higher non-surfactant nanoparticle volume fraction leads to a higher solid–liquid friction during the dynamic wetting process.

Figure 3.8 shows the effect of wettability of nanoparticles on the dynamic spreading of nanofluid droplets of $\varphi = 9.81$ %. As expected, the droplet spreading kinetics slows down when the interaction between particles and water molecules increases, which can be attributed to the increases in both the surface tension and the solid–liquid friction.

3.4 Local Dissipation Due to the Structural Disjoining Pressure

For a liquid pool or a thick liquid film, the liquid pressure equals the vapor pressure at the interface at equilibrium stage. However, for a thin liquid film, the solid–liquid interaction will become significantly strong when the liquid–vapor interface almost overlaps the solid–liquid interface. The strong solid–liquid interactions reduce the local pressure within the liquid film, resulting in an addition pressure difference between the liquid film and the bulk liquid. The additional pressure is referred to as disjoining pressure [45]. Therefore, the pressure of the liquid film is less than that of the vapor above the liquid ($p_L < p_V$). Electrostatic forces, the forces of "elastic" resistance of solvated, or adsorbed solvated, films, and the forces of molecular interaction can all contribute to the disjoining pressure. For an equilibrium thin film,

Fig. 3.9 Schematic of solid–
liquid–vapor molecules

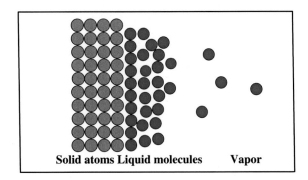

Solid atoms Liquid molecules   Vapor

the liquid pressure is $p_L = p_V + \Pi$, in which Π is disjoining pressure with a negative value. The negative disjoining pressure feed the thin film with liquid from the bulk phase. Generally, the disjoining pressure becomes significant when the liquid film thickness is less than several nanometers.

The relation of disjoining pressure Π and liquid film thickness δ_f can be derived from the interaction of the solid molecules and the liquid film molecules [46], as shown in Fig. 3.9.

The solid–liquid interaction can be described using LJ potential model

$$\phi_{SL}(r) = -\frac{C_{\phi,SL}}{r^3}\left[1 - \frac{D_m^6}{r^6}\right] \tag{3.6}$$

where ϕ_{SL} is the interaction potential of molecular pairs. $C_{\phi,SL}$ is the model constant. D_m is the minimal solid–liquid molecular distance, and r is the solid–liquid molecular distance. All the solid molecules were treated as a phase field. Integrated all the pairs interactions in the cylindrical coordinate obtains the interaction potential of solid surface and the thin liquid film, Φ_{SmL},

$$\Phi_{SmL} = -\frac{\pi\rho_S C_{\phi,SL}}{6r^3} + \frac{\pi\rho_S C_{\phi,SL}D_m^6}{45r^9} \tag{3.7}$$

where the ρ_S is the solid density. Substituting the Hamaker constant, A_{SL},

$$A_{SL} = \pi^2\rho_S\rho_L C_{\phi,SL} \tag{3.8}$$

into Eq. (3.7) gives the following equation:

$$\Phi_{SmL} = \frac{A_{SL}}{6\pi\rho_L D_m^6}\left[\frac{2}{15}\left(\frac{D_m}{r}\right)^9 - \left(\frac{D_m}{r}\right)^3\right], \tag{3.9}$$

where ρ_L is the liquid density.

Thus, the forces acting on the thin liquid film can be calculated from the gradient of the pair potential,

$$F_{\text{SL}} = -\nabla \Phi_{\text{SmL}} = \frac{A_{\text{SL}}}{2\pi\rho_{\text{L}}D_{\text{m}}^4} \left[\frac{2}{5}\left(\frac{D_{\text{m}}}{r}\right)^{10} - \left(\frac{D_{\text{m}}}{r}\right)^4 \right] r \qquad (3.10)$$

The pressure can be calculated from the Euler equation,

$$0 = -\frac{1}{\rho_{\text{L}}M/N_{\text{A}}}\nabla p + f_{\text{SL}} \qquad (3.11)$$

where M is the liquid molecular number, N_{A} is the Avogadro constant, f_{SL} is the force per unit mass on the liquid film.

Substituting Eq. (3.10) into (3.11) gives the following:

$$\frac{dp}{dr} = -\frac{A_{\text{SL}}}{2\pi D_{\text{m}}^4}\left[\frac{2}{5}\left(\frac{D_{\text{m}}}{r}\right)^{10} - \left(\frac{D_{\text{m}}}{r}\right)^4\right] \qquad (3.12)$$

Integrating Eq. (3.12) from the solid–surface interface to the liquid–vapor interface gives the following equation:

$$p_{\text{L,i}} = p_{\text{L,0}} + \frac{A_{\text{SL}}}{6\pi\delta_{\text{L}}^3} \qquad (3.13)$$

where $p_{\text{L,i}}$ is the internal pressure in the liquid film, while the $p_{\text{L,0}}$ is the pressure within the bulk liquid.

According to the definition of disjoining pressure,

$$\Pi = p_{\text{L,0}} - p_{\text{L,i}} = -\frac{A_{\text{SL}}}{6\pi\delta_f^3} \qquad (3.14)$$

where the Hamark constant, A_{SL}, depends on the liquid properties—for example, $A_{\text{SL}} = 1 \times 10^{-21}$ J for ammonium hydroxide. According to Eq. (3.14), the disjoining pressure increases with decreasing the thickness of thin film.

The disjoining pressure is extremely large within the ultrathin film, which significantly affects the flow fields in the junction of the thin film and the bulk liquid. Therefore, the disjoining pressure was widely used to explain various microfluidic problems, such as the thin film evaporation [47], microliquid layer in the liquid–vapor phase change [48], and precursor layer in the dynamic wetting [49].

The disjoining pressure can also be derived from thermodynamic aspect, as shown in Fig. 3.10. According to Gibbs–Duhem relation,

$$d\mu = -sdT + vdp \qquad (3.15)$$

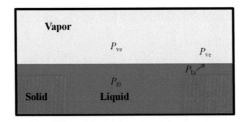

Fig. 3.10 Schematic of disjoining pressure within the thin liquid film

where μ is the chemical potential, s is the specific entropy, T is the temperature, v is the specific volume, and p is the pressure. Integrating the Gibbs–Duhem equation from the saturated state can obtain the chemical potential for the vapor phase,

$$\mu_V = \mu_{V,sat} + \int vdp, \tag{3.16}$$

and for the liquid phase,

$$\mu_L = \mu_{L,sat} + v_L(p_L - p_{sat}). \tag{3.17}$$

At the equilibrium stage, the vapor chemical equals the liquid chemical potential at the interface,

$$\mu_V = \mu_L. \tag{3.18}$$

However, for the thin liquid film, the liquid chemical potential contains additional potential due to the attraction of solid surface molecules, Φ_{SmL}, which is given as follows:

$$\mu_L = \mu_{L,sat} + v_L(p_L - p_{sat}) + \Phi_{SmL}. \tag{3.19}$$

At the liquid–vapor interface,

$$p_{L,i} = p_V. \tag{3.20}$$

Substituting Eqs. (3.16) and (3.19) into (3.18), we have

$$\int vdp = v_L(p_L - p_{sat}) + \Phi_{SmL} \tag{3.21}$$

For the ideal gas assumption, the left-hand term in Eq. (3.21) can be calculated as follows:

$$RT \ln\left(\frac{p_V}{p_{sat}}\right) = v_L(p_L - p_{sat}) + \Phi_{smf}, \tag{3.22}$$

Fig. 3.11 Comparison of the present molecular simulation with the classical theory and the previous molecular dynamic simulation results [46]

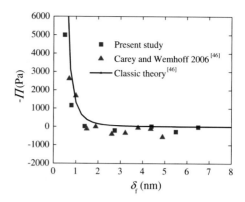

which can be rewritten as

$$\frac{p_V}{p_{sat}} = \exp\left\{\frac{p_{sat}v_L}{RT}\left(\frac{p_V}{p_{sat}} - 1\right) + \frac{\Phi_{SmL}}{RT}\right\} \qquad (3.23)$$

since $p_V \approx p_{sat}$, the first term at the right-hand side can be omitted. Therefore, we can obtain the vapor above the thin film by substituting the Eq. (3.9) into (3.23) without considering the r^{-9} terms,

$$\frac{p_V}{p_{sat}} = \exp\left\{-\frac{A_{SL}}{6\pi\rho_L\delta_L^3 k_B T}\right\} \qquad (3.24)$$

in which the thin liquid film thickness D_m was replaced by δ_L.

Substituting Eqs. (3.14) into (3.24) gives the following equation:

$$\Pi = \rho_L k_B T \ln\left(\frac{p_V}{p_{sat}}\right) \qquad (3.25)$$

Due to the molecular instinct of the thin film, we calculate the disjoining pressure using the MD simulations. The disjoining pressure was calculated by the definition of $\Pi = p_b - p_{thin}$, in which p_b is the pressure of bulk phase, p_{thin} is the pressure in the thin film. In addition, the disjoining can also be calculated by Eq. (3.24) as a comparison. Figure 3.11 shows the comparison of the present molecular simulation with the classical theory [46] and the previous MD simulation results [46] for an Argon thin film on the gold solid surface. The results show that the present simulation agrees well with the classical theory.

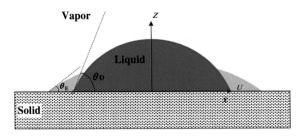

Fig. 3.12 Schematic of the contact line motion for a spreading droplet

As shown in Fig. 3.12, for a spreading droplet, the contact line motion is driven by the unbalanced Young's force,

$$F = L(\sigma_{SV} - \sigma_{SL} - \sigma_{LV} \cos \theta_D) \tag{3.26}$$

where σ is the surface tension, and the subscripts S, L, and V donate solid, liquid, and vapor, respectively. L is the contact line length. θ_D is the dynamic contact angle, and θ_E is the equilibrium contact angle. For the complete wetting cases, the θ_E equals zero. Equation (3.26) can be rewritten as follows using Young's equation.

$$F = \sigma_{LV}L(\cos \theta_E - \cos \theta_D) \tag{3.27}$$

For the complete wetting cases, the driven force is given as,

$$F = L(\sigma_{SV} - \gamma_{SL} - \gamma_{LV}) = LS \tag{3.28}$$

where S is the spreading coefficient.

For the wedge shape of the contact line region, the thickness of the drop decreases from the bulk phase to the contact line region. The reducing of thickness will result in the pressure deference between the contact line region and the bulk droplet. The pressure deference leads to the disjoining pressure which can also trigger the contact line motion. What is the main driving force for the contact line motion, the unbalanced Young's stress or the disjoining pressure? The answer to this question may lead to the fundamental understanding of the contact line motion mechanism. For nanofluids, as discussed in Sect. 3.3, the ordering solid-like nanostructure in the vicinity of the contact line region results in the additional disjoining pressure, referred as to structural disjoining pressure, facilitating the dynamic wetting of nanofluids. The simulations were performed in a NVT ensemble with 13,500 water molecules. To examine the roles of structural disjoining pressure in the nanofluid dynamic wetting, several self-assembled gold nanoparticles with diameter of 3.26 nm were deposited on the gold solid surface. Thus, the thickness of thin liquid film above the deposited nanoparticle (d_1 = 2.21 nm) is less than that of the thin liquid film above the solid surface without nanoparticles (d_2 = 5.47 nm), as shown in Fig. 3.13. Correspondingly, the local pressure of the thin liquid film above the nanoparticles (regions 1 and 2,

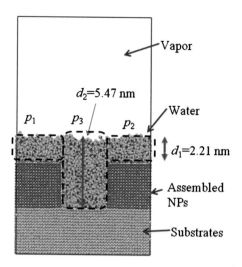

Fig. 3.13 Thickness of liquid film variety due to the nanoparticle self-assembly

$\bar{p}_1 = \bar{p}_2$) is less than that of the thin liquid film above the solid surface without nanoparticle (region 3, $\bar{p}_3$), leading to an additional pressure gradient which can drive the thin film motion, as shown in Fig. 3.14. This additional pressure gradient due to the nanoparticle self-assembly is referred to as structural disjoining pressure. For a thin liquid film with a cross-sectional area of $A = 3.2 \times 2.2$ nm^2, the driven force due to the disjoining pressure can be calculated approximately as follows:

$$F_{\text{sdp}} = \Pi A = (\bar{p}_1 - \bar{p}_3)A \approx -3.5 \times 10^{-9} \, \text{N} \tag{3.29}$$

For a complete wetting thin liquid film ($\theta_E = 0$) with a wetting contact angle of $\theta_D = 90°$, as shown in Fig. 3.15, the unbalanced Young's stress acting on the thin film can be calculated by the following:

$$F_{\text{st}} = L\sigma_{\text{LV}}(\cos \theta_E - \cos \theta_D) \approx 3.4 \times 10^{-10} \, \text{N}, \tag{3.30}$$

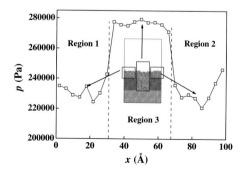

Fig. 3.14 Local pressure at various locations with different thicknesses of a thin liquid film

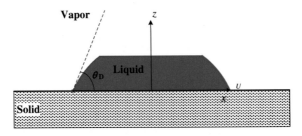

Fig. 3.15 Schematic of Young's unbalance force for a thin liquid film motion

which is 1/10 of the driving force due to the structural disjoining pressure calculated by Eq. (3.29). Therefore, the structural disjoining pressure due to the nanoparticle self-assembly facilitates the contact line motion.

3.5 Conclusions

In this chapter, the dynamic spreading of water nanodroplets containing nanoparticles and the effects of structural disjoining pressure are examined via MD simulations. The main conclusions are as follows:

1. The nanoparticle diffusion time is larger than the nanosize droplet spreading time. The nanoparticles do not have enough time to diffuse to the vicinity of the contact line region, thus the self-assembly of nanoparticles does not occur. The addition of non-surfactant nanoparticles hinders rather than enhances the droplet spreading kinetics during the nanosecond process.
2. The contact line velocity decreases with increasing nanoparticle volume fraction and particle–water interactions, as a result of increasing surface tension and solid–liquid friction and the absence of nanoparticle ordering in the vicinity of the contact line.
3. The structural disjoining pressure is ten times higher than the unbalanced Young's stress, which can facilitate the contact line motion if the nanoparticle self-assembly occurs in the vicinity of the contact line region.

References

1. Wasan DT, Nikolov AD (2003) Spreading of nanofluids on solids. Nature 423:156–159
2. Chaudbury MJ (2003) Spread the word about nanofluids. Nature 423:131–132
3. Kondiparty K, Nikolov AD, Wu S et al (2011) Wetting and spreading of nanofluids on solid surfaces driven by the structural disjoining pressure: statics analysis and experiments. Langmuir 27:3324–3335

4. Kondiparty K, NikolovAD Wasan DT et al (2012) Dynamic spreading of nanofluids on solids part I: experimental. Langmuir 28:14618–14623
5. Liu KL, Kondiparty K, Nikolov AD et al (2012) Dynamic spreading of nanofluids on solids part II: modeling. Langmuir 28:16274–16284
6. Sefiane K, Skilling J, MacGillivary J (2008) Contact line motion and dynamic wetting of nanofluid solutions. Adv Colloid Interface Sci 138:101–120
7. Shen J, Liburdy JA, Pence DV et al (2009) Droplet impingement dynamics: effect of surface temperature during boiling and non-boiling conditions. J Phys: Condens Matter 21:464133
8. Radiom M, Yang C, Chan WK (2013) Dynamic contact angle of water-based titanium oxide nanofluid. Nanoscale Res Lett 8:282
9. Sefiane K, Bennacer R (2009) Nanofluids droplets evaporation kinetics and wetting dynamics on rough heated substrates. Adv Colloid Interface Sci 147–148:263–271
10. Moffat JR, Sefiane K, Shanahan MER (2009) Effect of TiO_2 nanoparticles on contact line stick-slip behavior of volatile drops. J Phys Chem B 113:8860–8866
11. Murshed SMS, Nieto de Castro CA, Lourenco MJV et al (2007) A review of boiling and convective heat transfer with nanofluids. Renew Sust Energ Rev 15:2342–2354
12. Wen DS (2008) Mechanisms of thermal nanofluids on enhanced critical heat flux (CHF). Int J Heat Mass Transfer 51:4958–4965
13. Kim SJ, Bang IC, Buongiorno J, Hu LW (2006) Effects of nanoparticle deposition on surface wettability influencing boiling heat transfer in nanofluids. Appl Phys Lett 89:153107
14. Sefiane K (2006) On the role of structural disjoining pressure and contact line pinning in critical heat flux enhancement during boiling of nanofluids. Appl Phys Lett 89:044106
15. Cheng SF, Grest GS (2012) Structure and diffusion of nanoparticle monolayers floating at liquid/vapor interfaces: a molecular dynamics study. Soft Condensed Matte 136:214702
16. Frost DS, Dai LL (2011) Molecular dynamics simulations of nanoparticle self-assembly at ionic liquid-water and ionic liquid-oil interfaces. Langmuir 27:11339–11346
17. Luo MX, Dai LL (2007) Molecular dynamics simulations of surfactant and nanoparticle self-assembly at liquid-liquid interfaces. J Phys: Condens Matter 19:375109
18. Luo M, Mazyar OA, Zhu Q et al (2006) Molecular dynamics simulation of nanoparticle self-assembly at a liquid-liquid interface. Langmuir 22:6385–6390
19. Song Y, Luo M, Dai LL (2010) Understanding nanoparticle diffusion and exploring interfacial nanorheology using molecular dynamics simulations. Langmuir 26:5–9
20. Frost DS, Dai LL (2012) Molecular dynamics simulations of charged nanoparticle self-assembly at ionic liquid-water and ionic liquid-oil interfaces. J Chem Phys 136:084706
21. Drazer G, Koplik J, Acrivos A et al (2002) Adsorption phenomena in the transport of a colloidal particle through a nanochannel containing a partially wetting fluid. Phys Rev Lett 89:244501
22. Drazer G, Khusid B, Koplik J et al (2005) Wetting and particle adsorption in nanoflows. Phys Fluids 17:017102
23. Li ZG, Drazer G (2008) Hydrodynamic interactions in dissipative particle dynamics. Phys Fluids 20:103601
24. Kim HS, Park SS, Hagelberg F (2011) Computational approach to drying a nanoparticle-suspended liquid droplet. J Nanopart Res 13:59–68
25. Horn HW, Swope WC, Pitera JW et al (2004) Head-Gordon T. Development of an improved four-site water model for biomolecular simulations: TIP4P-Ew. J Chem Phys 120:9665
26. Ryckaert JP, Ciccotti G, Berendsen HJC (1977) Numerical integration of the cartesian equations of motion of a system with constraints: molecular dynamics of n-alkanes. J Computl Phys 23:327–341
27. Daw MS, Foiles SM, Baskes MI (1993) The embedded-atom method: a review of theory and applications. Mat Sci R 9:251–310
28. Schravendijk P, van der Vegt N, Delle Site L, Kremer K et al (2005) Dual-scale modeling of benzene adsorption onto Ni(111) and Au(111) surfaces in explicit water. Chem Phys Chem 6:1866–1871

29. Plimpton S (1995) Fast parallel algorithms for short-range molecular dynamics. J Chem Phys 117:1–19
30. Tanner LH (1979) The spreading of silicone oil drops on horizontal surfaces. J Phys D 12:1473–1484
31. Blake TD, Haynes JM (1969) Kinetics of liquid/liquid displacement. J Colloid Interface Sci 30:421–423
32. Blake TD (2006) The physics of moving wetting lines. J Colloid Interface Sci 299:1–13
33. Lu G, Duan YY, Wang XD (2015) Effects of free surface evaporation on water nanodroplet wetting kinetics: a molecular dynamics study. J Heat Transfer 137:091001–091006
34. Einstein A (1905) Über die von der molecular kinetischen Theorie der Wärmege forderte Bewegung von in ruhenden Flüssigkeiten suspendierten Teilchen. Ann Phys 322:541–549
35. Liang ZP, Wang XD, Lee DJ et al (2009) Spreading dynamics of power-law fluid droplets. J Phys Condens Matter 21:464117
36. Wang XD, Zhang Y, Lee DJ et al (2007) Spreading of completely wetting or partially wetting power-law fluid on solid surface. Langmuir 23:9258–9262
37. Lu G, Hu H, Duan YY, Sun Y (2013) Wetting kinetics of water nano-droplet containing non-surfactant nanoparticles: a molecular dynamics study. Appl Phys Lett 103:253104
38. Zosel A (1993) Studies of the wetting kinetics of liquid drops on solid surfaces. Prog Colloid Polym Sci 271:680–687
39. Ismail AE, Grest GS, Stevens MJ (2006) Capillary waves at the liquid-vapor interface and the surface tension of water. J Chem Phys 125:014702
40. Gittens GJ (1969) Variation of surface tension of water with temperature. J Colloid Interface Sci 30:406–412
41. Tanvir S, Li Q (2012) Surface tension of Nanofluid-type fuels containing suspended nanomaterials. Nanoscale Res Lett 7:226
42. Moosavi M, Goharshadi EK, Youssefi A (2010) Fabrication, characterization, and measurement of some physicochemical properties of ZnO nanofluids. Int J Heat Fluid Flow 31:599–605
43. Kwok DY, Neumann AW (2000) Contact angle interpretation in terms of solid surface tension. Colloids Surf A 161:31–49
44. Chen RH, Phuoc TX, Martello D (2011) Surface tension of evaporating nanofluid droplets. Int J Heat Mass Transfer 54:2459–2466
45. Israelachvili JN (2011) Intermolecular and Surface Forces. Academic Press, Singapore
46. Carry VP, Wemhoff AP (2006) Disjoining pressure effects in ultra-thin liquid films in micropassages—comparison of thermodynamic theory with predictions of molecular dynamics simulations. ASME J Heat Transfer 128:1276–1284
47. Ma HB, Cheng P, Borgmeyer B (2008) Fluid flow and heat transfer in the evaporating thin film region. Microfluidic Nanofluidic 4:237–243
48. Katto Y, Shoji M (1970) Principal mechanism of micro-liquid-layer formation on a solid surface with a growing bubble in nucleate boiling. Int J Heat Mass Transfer 13:1299
49. Craster RV, Matar OK (2009) Dynamics and stability of thin liquid films. Rev Mod Phys 81:1131–1198

Chapter 4
Bulk Dissipation in Nanofluid Dynamic Wetting: Wettability-Related Parameters

Abstract In this chapter, we study how nanoparticles alter the surface tension, viscosity, and rheology of nanofluids from microscopic viewpoints using molecular dynamics simulations. The results reveal the roles of additional nanoparticles on the modification of wettability-related parameters (surface tension, viscosity, and rheology) and then provide the guidelines in building nanofluid dynamic wetting models.

4.1 Introduction

The suspensions of nanoparticles in the nanofluids significantly modify the properties of the base fluids. Therefore, nanofluids exhibit attractive properties, such as high thermal conductivity and tunable surface tension, viscosity, and rheology. Various attempts have been made to understand the mechanisms for these property modifications caused by the additional nanoparticles. However, these mechanisms are still unclear due to the lack of direct nanoscale evidences.

Most previous studies on the thermophysical properties of nanofluids have explored the effects of the nanofluid parameters, such as the nanoparticle loading, diameter, and material and the base fluid type on the thermal conductivity, surface tension, viscosity, and rheology. Among these properties, the surface tension, viscosity, and rheology are related to the dynamic wetting, a process dominated by the surface tension and viscous forces. Therefore, the surface tension, viscosity, and rheology are defined as wettability-related parameters in this book. The modification of surface tension, viscosity, and rheology by adding nanoparticles greatly affects the dynamic wetting behaviors. These effects are regarded as the bulk dissipation here. Nanofluids are reported to have higher thermal conductivity than the base fluids, which were explained by several established models, e.g., Brownian motion, the solid/liquid interface layer around the nanoparticles, or the ballistic phonon transport hypotheses [1–10]. There have been many studies on thermal conductivity, but only a few studies on nanofluid surface tension, viscosity, and

© Springer-Verlag Berlin Heidelberg 2016
G. Lu, *Dynamic Wetting by Nanofluids*,
Springer Theses, DOI 10.1007/978-3-662-48765-5_4

Table 4.1 Studies of the surface tension of nanofluids

Authors	Nanofluids	Method	σ
Zhu et al. [12]	Al_2O_3/H_2O	Experiment	Increase
Moosavi et al. [13]	ZnO/EG/glycerol	Experiment	Increase
Tanvir and Li [14]	Al/Al_2O_3/B/MWCNT H_2O	Experiment	Increase
Kumar and Milanova [15]	CNT/H_2O	Experiment	Increase
Chen et al. [16]	Laponite/Fe_2O_3/ Ag/H_2O	Experiment	Constant/decrease
Das et al. [17]	Al_2O_3/H_2O	Experiment	Constant
Vafaei et al. [18]	Bi_2Te_3/AOT	Experiment/model	Decrease/increase
Murshed et al. [19]	TiO_2/H_2O	Experiment	Decrease
Radiom et al. [20]	TiO_2/H_2O	Experiment	Decrease
Liu and Kai [21]	TiN/SiC/Al_2O_3/ CNT/ammonia–H_2O	Experiment	TiN/SiC decrease Al_2O_3/CN increase

Reprinted from Ref. [11], with kind permission from Springer Science+Business Media

rheology. Table 4.1 summarizes the studies on the nanofluid surface tensions [11]. The studies are still controversial because the surface tensions of nanofluids have been reported to be increased [12–14], unchanged [15, 16], or decreased [17, 18] compared with that of the base fluids. Even for the same nanofluids, such as Al_2O_3/ H_2O nanofluids, surface tension was reported to be increased by Zhu et al. [12], but unchanged by Das et al. [17]. Some studies have tried to qualitatively explain the mechanisms of these modifications in the nanofluid surface tension [14, 19–21]. Tanvir and Li [14] suggested that the attractive forces between the particles at the liquid–vapor interface increase as the particle concentration increases which increase the surface tension. However, Murshed et al. [19], Radiom et al. [20], and Liu and Kai [21] proposed that the surface tension reduction may be attributed to the reduction of the cohesive energy at the liquid–vapor interface, because nano-sized particles are brought to the lowest interfacial energy level by the Brownian motion. They also suggested that the nanoparticles function as surfactant molecules since nanoparticles are absorbed onto the liquid–vapor interface to reduce the surface tension. However, all these analyses are only suggestions without any direct evidences, and some are even contradictory.

The nanofluids are usually regarded as colloidal suspensions. Rigorously speaking, colloidal suspensions are mixture solutions containing micro-/milli-sized particles, while nanofluids containing nanosized particles. Several classical models have been proposed to predict the effective viscosity of colloidal suspensions [22–25]. However, these models failed to predict the effective viscosity of nanofluids [26–28]. The models of nanofluid viscosity have been reviewed by Eastman et al. [29], Keblinski et al. [30], and Mahbubul et al. [31]. Most of these models were only based on hypotheses or empirical correlations between the effective viscosity and macroscopic parameters, such as the loading fractions or temperatures. Some

Table 4.2 Studies of the rheology of nanofluids

Authors	Nanofluids	Method	Rheology
Das et al. [17]	Al_2O_3/H_2O, CuO/H_2O	Experiment	Newtonian
Prasher et al. [35]	Al_2O_3/PG	Experiment	Newtonian
Chen et al. [36]	TiO_2/ethylene glycol	Experiment	Newtonian
Susan-Resiga et al. [37]	Fe_3O_4/MOL	Experiment	Newtonian
Wang et al. [38]	CuO/H_2O/ethylene glycol, Al_2O_3/H_2O/ethylene glycol/engine oil	Experiment	Non-Newtonian
He et al. [39]	TiO_2/H_2O	Experiment	Non-Newtonian
Murshed et al. [40]	Al_2O_3/H_2O	Experiment	Non-Newtonian
Chen et al. [41]	TNT/EG	Experiment	Non-Newtonian (shear thinning)
Ding et al. [42]	CNT/H_2O	Experiment	Non-Newtonian (shear thinning)
Kole and Dey [43]	Al_2O_3/car coolant, CuO/gear oil	Experiment	Non-Newtonian
Yu et al. [44]	AlN/Eg, AlN/PG	Experiment	$\varphi < 5\ \%$: Newtonian, $\varphi > 5\ \%$: non-Newtonian
Yu et al. [45]	ZnO/EG	Experiment	$\varphi < 2\ \%$: Newtonian, $\varphi > 3\ \%$: non-Newtonian
Kim et al. [46]	Al_2O_3/H_2O	Experiment	$\varphi < 2\ \%$: Newtonian, $3\ \% < \varphi < 5\ \%$: non-Newtonian
Abareshi et al. [47]	α-Fe_2O_3-glycerol	Experiment	Non-Newtonian@low temperature

Reprinted from Ref. [11], with kind permission from Springer Science+Business Media

correlations had considered the effects of particle size, which is related to the Brownian motion [32, 33]; however, undetermined constants in these models must be fit from macroscopic experimental measurements. Thus, more studies are needed to identify the enhancement mechanism with microscopic evidence.

The rheological properties of nanofluids had been reviewed by Chen et al. [34]. According to the summary in Table 4.2, there is still debate about whether nanofluids exhibit Newtonian [16, 35–37] or non-Newtonian [38–47] in experiments. Even for the same nanofluid, such as Al_2O_3/H_2O nanofluids, Newtonian rheological behavior was reported by Das et al. [17] and Prasher et al. [35], while non-Newtonian rheological behavior was observed by Wang et al. [38], Murshed et al. [40], Kole and Dey [43], and Kim et al. [46]. Most of these studies mainly reported the experimental data, with few focused on the mechanisms for the rheological transformation from Newtonian to non-Newtonian fluids by adding nanoparticles. Yu et al. [45] stated that the rheological properties of nanofluids depend strongly on many factors, such as nanoparticle material, shape, the loading fractions, and temperatures. Chen et al. [34] proposed an aggregation mechanism to interpret the rheological behavior of nanofluids and categorized the rheological behavior of nanofluids into four groups such as dilute, semi-dilute, semi-concentrated, and concentrated. However, they did not consider the mechanism of rheological transform. In addition, no direct evidence was provided to support their hypotheses.

Most previous studies of nanofluid have focused on collecting the thermophysical property data. The macroscopic measurements were then used to establish models to explain the mechanisms; however, these explanations have not been supported by direct evidence. MD simulations can provide microscopic understanding of the thermophysical properties in much greater detail, which have been extensively used to investigate the surface tension, viscosity, and rheology of pure liquids [48–55] or binary mixture solutions [56–60]. Recently, the MD simulations have also been used to model the nanofluid thermal conductivity [61–67] with some promising results that verify the theories based on macroscopic measurements. However, few MD studies have focused on the surface tension, viscosity, and rheology of nanofluids.

The objective of this chapter is to study how nanoparticles alter the surface tension, viscosity, and rheology of nanofluids from microscopic viewpoints using MD simulations. The results reveal the roles of additional nanoparticles on the modification of wettability-related parameters and then provide the guidelines in building nanofluid dynamic wetting models.

4.2 Simulation Methods

4.2.1 Simulation Systems

Figure 4.1 shows the simulated systems had four nanofluid loadings. Bulk water films (4500 water molecules) were simulated at 300 K using the molecular dynamic simulations with four gold nanoparticle loadings (φ = 0, 3.43 %, 6.77 %, and

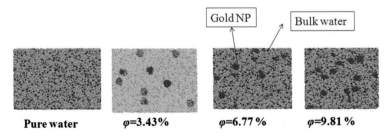

Pure water φ=3.43% φ=6.77% φ=9.81%

Fig. 4.1 Illustrations of the molecular structures of pure water and three gold nanoparticle nanofluid loading fractions (the water molecules are transparent to illustrate the nanoparticle positions for φ = 3.43 %). Reprinted from Ref. [11], with kind permission from Springer Science+Business Media

9.81 %). The gold nanoparticles ($0.8 \times 0.8 \times 0.8$ nm^3) randomly distributed inside the bulk liquid. The four-point TIP4P-Ew water model, PPPM technique, and SHAKE algorithm were used to describe the water–water interactions [68, 69]. The EAM [70] was used for the gold–gold interactions. The simulation details have been described in Chap. 3. A 12-6 LJ potential with $\sigma = 3.1$ Å [71] and a cutoff distance of 9 Å was used to describe the water–gold interactions. The gold nanoparticle wettability was changed with different gold–water interaction parameter ε ($\varepsilon = 0.0070, 0.02714, 0.05427$ and 0.08141 eV). The *NVT* ensembles at $T = 300$ K with a time step of 1 fs were used to calculate the thermodynamic properties. The simulations were performed with LAMMPS software packages [72].

4.2.2 Surface Tension Calculation Method

In most MD simulations, the surface tension is calculated based on Young–Laplace equation [73], as shown in Eq. (4.1).

$$\Delta p = \frac{\sigma_{LV}}{R} \tag{4.1}$$

where Δp is the pressure difference between the inside and outside of the droplet, σ_{LV} is the liquid–vapor surface tension, and R is the droplet radius. However, the pressure fluctuates greatly in LAMMPS. For example, for a set pressure of 1 bar, the standard deviation of the fluctuations is about 40 bar. Thus, the surface tension cannot be calculated with such large pressure variations even with time averaging. Therefore, a new method, referred to as excess energy method here, was used to calculate the surface tension. The simulation details are shown in Chap. 3.

4.2.3 Viscosity and Rheology Calculation Method

The reverse non-equilibrium MD (rNEMD) method [74] was used to calculate the nanofluid viscosity. The method is based on Onsager linear response theory,

$$j_\alpha = - \sum_\beta L_{\alpha\beta} X_\beta \tag{4.2}$$

where j_α is the momentum flux, X_β is the driving force (velocity gradient), and $L_{\alpha\beta}$ is the diffusion coefficient (viscosity).

As shown in Fig. 4.2, a momentum flux was imposed on the bulk liquid to generate a velocity gradient. The diffusion coefficient, in this case the viscosity, was then obtained using:

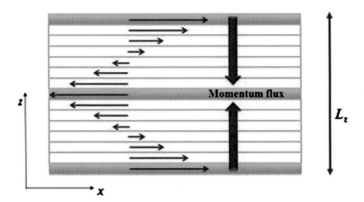

Fig. 4.2 Momentum exchange in the rNEMD simulation method. Reprinted from Ref. [11], with kind permission from Springer Science+Business Media

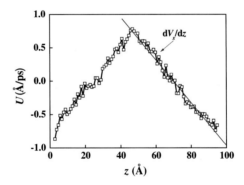

Fig. 4.3 Velocity profile in the rNEMD method. Reprinted from Ref. [11], with kind permission from Springer Science+Business Media

$$j_\alpha \langle p_x \rangle = \frac{P_x}{2tA} \tag{4.3}$$

where P_x is the pressure in the x direction, t is the velocity swapping period, and A is the area of xy plane.

To impose the momentum flux onto the bulk liquid, two water molecules were moved in the bottom and middle plates against the intended current, one with the minimum velocity and the other with the maximum velocity. The velocities were then swapped in another calculation. The momentum flux then generated a velocity gradient, dV_x/dz. As shown in Fig. 4.3, a good linear velocity profile was obtained along the z direction that was fit with a straight line with the momentum flux and then calculated using Eq. (4.3).

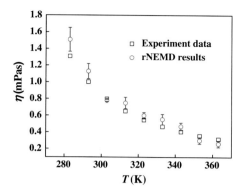

Fig. 4.4 Verification of the rNEMD method: viscosity of pure water at various temperatures. Reprinted from Ref. [11], with kind permission from Springer Science+Business Media

The velocity swapping period, t, in Eq. (4.3) was varied to create various momentum fluxes and velocities to relate the viscosity to the shear rate (rheology).

Figure 4.4 compares the present MD results with the experimental data for pure water at various temperatures. The good agreement indicates the accuracy of the present water model and the rNEMD method.

4.3   Results and Discussion

4.3.1   Surface Tension of Gold–Water Nanofluids

For pure water, the calculated surface tension from the MD simulations was $\sigma = 0.0679$ N/m at $T = 300$ K, close to the experimental value of $\sigma = 0.072$ N/m [75]. For the 3.43 % nanofluid, the addition of nanoparticles increased the surface tension for the nanoparticles with $\varepsilon_{water–gold} = 0.05427$ eV, but reduced it for those with $\varepsilon_{water–gold} = 0.0070$ eV. The different tendencies for these two cases are related to the different water–gold interactions ($\varepsilon_{water–gold}$) and water–water interactions ($\varepsilon_{water–water} = 0.0071$ eV for TIP4P-Ew). For $\varepsilon_{water–gold} < \varepsilon_{water–water}$, the nanoparticles are hydrophobic, so they tend to stay on the free surface (liquid–vapor interface), acting as surfactant-like particles, as shown in Fig. 4.5a. For $\varepsilon_{water–gold} > \varepsilon_{water–water}$, the nanoparticles are hydrophilic, so they tend to submerge into the bulk liquid, acting as non-surfactant particles, as shown in Fig. 4.5b.

The surface tension indicates the unbalanced forces acting on the liquid molecules at the interface due to the van der Waals force, as shown in Fig. 4.6a. The liquid molecules at the surface do not have an equal number of molecules in the vapor side, so they are pulled inward by the internal molecules, which results in the surface tension, pulling the liquid surface to contract to the minimum area. Thus, the nanoparticle wettability is responsible for the different surface tensions. For

Fig. 4.5 Snapshots of the
3.43 % nanofluid with
a hydrophobic gold
nanoparticles ($\varepsilon = 0.007$ eV);
b hydrophilic gold
nanoparticles
($\varepsilon = 0.05427$ eV). Reprinted
from Ref. [11], with kind
permission from Springer
Science+Business Media

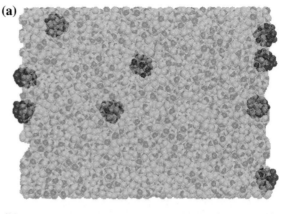

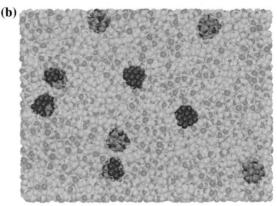

hydrophobic nanoparticles, the nanoparticles gather on the free surface. The repulsion force between the nanoparticles and the water molecules increases the intermolecular spacing at the interface and reduces the attraction forces between the water molecules inside the bulk liquid and the ones on the free surface regions, thus reducing the surface tension, as shown in Fig. 4.6b. However, some hydrophilic nanoparticles are transported toward the interfacial region by Brownian motion where the attraction forces between the nanoparticles and the water molecules reduce the intermolecular spacing at the interface. The water molecules at the free surface are more strongly pulled inward due to the presence of the hydrophilic nanoparticles with stronger gold–water interaction forces than those with the water–water interactions, which increase the surface tension, as shown in Fig. 4.6c.

Figure 4.7 shows the evidences for the hydrophobic/hydrophilic nanoparticles changing the intermolecular spacing in the interfacial region. The water density remains constant in the bulk liquid but decreases sharply near the interface. The thickness of the reduced density region, defined as the interface width, d, decreases with increasing nanoparticle wettability, leading to the increased surface tension. In

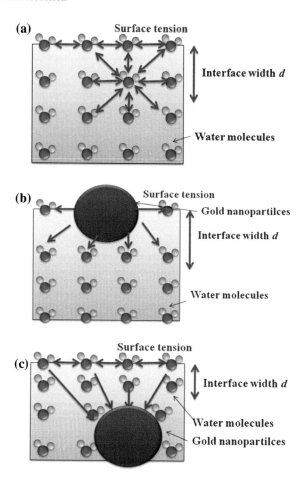

Fig. 4.6 Schematic of the van der Waals forces between water molecules and nanoparticle–water molecule near the interfacial region: **a** water molecular interaction for pure water; **b** gold nanoparticle–water molecule interactions for $\varepsilon_{water-gold} < \varepsilon_{water-water}$; **c** gold nanoparticle–water molecule interactions for $\varepsilon_{water-gold} < \varepsilon_{water-water}$. Reprinted from Ref. [11], with kind permission from Springer Science+Business Media

addition, the interface width, d, decreases with increasing φ for the hydrophilic nanoparticles, but increases with increasing φ for the hydrophobic nanoparticles.

Therefore, for nanofluids, the dynamic wetting can be facilitated by adding hydrophobic nanoparticles, but inhibited by adding hydrophilic nanoparticles, which is related to the modification of nanofluid surface tension.

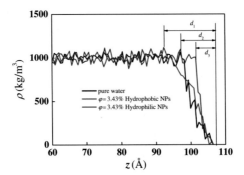

Fig. 4.7 Water density profiles perpendicular to the free surface for pure water and nanofluids with hydrophobic or hydrophilic nanoparticles ($\varphi = 3.43$ %). Reprinted from Ref. [11], with kind permission from Springer Science+Business Media

4.3.2 Viscosity of Gold–Water Nanofluids

Figure 4.8 shows the viscosities of pure water and low loading nanofluids ($\varphi = 0$ %, $\varphi = 3.43$ %, and $\varphi = 6.77$ %). The viscosity increases with increasing volume concentration, which agrees with experimental data [76]. The increasing viscosity with loading is illustrated by the microscopic picture of one randomly selected gold nanoparticle shown in the inset in Fig. 4.9, in which the hydrogen atoms are hidden. An absorbed water layer has formed around the gold nanoparticle, which can be observed by plotting the water density along the radial direction, as shown in Fig. 4.9. The water density near the nanoparticle at 19 Å is five times that of the bulk liquid, indicating a solid-like absorbed water layer around the gold nanoparticle. The effect can be explained by the Einstein diffusion equation [77],

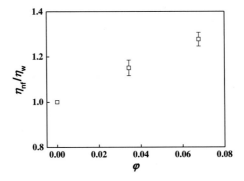

Fig. 4.8 Nanofluid viscosity for various NP loadings. Reprinted from Ref. [11], with kind permission from Springer Science+Business Media

Fig. 4.9 Water density near a gold nanoparticle (*Inset* illustration of the water molecule layer). Reprinted from Ref. [11], with kind permission from Springer Science+Business Media

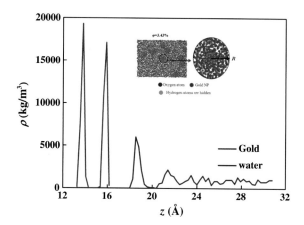

$$D_{\mathrm{NP}} = \frac{k_{\mathrm{B}}T}{6\pi\mu r} \tag{4.4}$$

where D_{NP} is the nanoparticle diffusion coefficient, k_{B} is Boltzmann's constant, T is the absolute temperature, μ is the base liquid viscosity, and r is the particle radius. The absorbed water layer increases the equivalent nanoparticle radius which hinders nanoparticle diffusion within the base liquid according to Eq. (4.4). Consequently, the increased nanofluid viscosity can be explained by the decreased nanoparticle diffusion coefficient. The absorbed layer is also reported as the reasons for the nanofluid thermal conductivity enhancement [51, 53]. The interactions between the nanoparticles and the water molecules were changed by modifying ε. As shown in Fig. 4.10a, the density of the absorbed water layer increases with the increasing gold–water interaction parameter. Thus, smaller diffusion coefficients occur for stronger NP–water molecule interactions, and the nanofluid viscosity can be expected to increase with increasing the gold–water interaction parameter, as shown in Fig. 4.10b. The number of gold nanoparticles in the solution affects the viscosity, as shown in Fig. 4.10b, but does not affect the intensity of the absorbed water layer for a single nanoparticle if the nanoparticle loading is low enough.

4.3.3   Rheology of Gold–Water Nanofluids

Figure 4.11 shows the viscosity versus shear rate variation, as known as rheological relations. For pure water and low nanofluid loadings ($\varphi = 3.43$ % and $\varphi = 6.77$ %), the viscosity remains constant for shear rates from 10^{-1} to 3×10^{2} s^{-1}. However, the viscosity of the $\varphi = 9.81$ % nanofluid remains constant only with a narrow shear rate range (from 10^{-1} to 10^{0} s^{-1}) and decreases as the shear rate increases from 10^{0} to 3×10^{2} s^{-1}, indicating shear-thinning non-Newtonian rheological behavior. This

Fig. 4.10 Effects of gold
nanoparticle–water molecule
interactions on **a** atom number
density; **b** viscosity.
Reprinted from Ref. [11],
with kind permission from
Springer Science+Business
Media

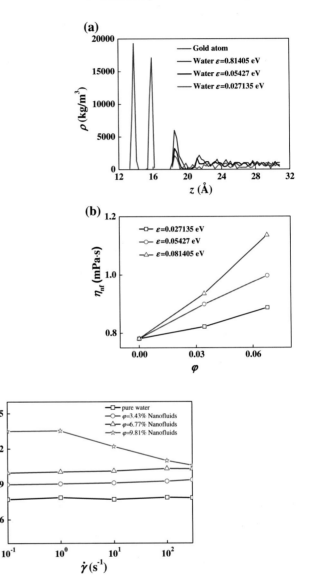

Fig. 4.11 Viscosity versus shear rate (rheology) for nanofluids for various loadings
($\varepsilon = 0.05427$ eV). Reprinted from Ref. [11], with kind permission from Springer Science+Business
Media

rheology transformation at high shear rates was also seen experimentally by Carré
and Woehl [78].

The inset in Fig. 4.12 shows a typical microscopic structure of two nanoparticle–
water molecule cluster. Comparison of the particle distribution with the high
loading in the inset in Fig. 4.12 with the distribution for a lower loadings seen in the

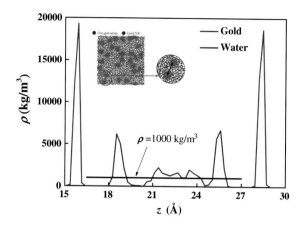

Fig. 4.12 Solidification effect at high loadings (*Inset* illustration of the molecular spacing). Reprinted from Ref. [11], with kind permission from Springer Science+Business Media

inset in Fig. 4.9 shows that there are more chances for nanoparticles to move closer together for high loadings since nanoparticles always move randomly in the base liquid due to Brownian motion. The solid-like absorbed water layers around the nanoparticles are also observed in this structure. Additionally, the water density between these two closed nanoparticles is several times larger than that of the bulk liquid density, as shown in Fig. 4.12, indicating that solidification of the water molecules also occurs in this structure, which is defined here as a solidification structure due to the additional nanoparticle–nanoparticle interactions. This solidification structure further increases the nanofluid viscosity for higher loadings. The solidification structures remain unchanged at low shear rates, where the nanofluids exhibit Newtonian rheological behavior. However, there is a critical shear rate above which the solidification structures are disrupted and the solidified water molecules move more freely, which reduces the nanofluid viscosity, leading to the shear-thinning non-Newtonian rheological behavior.

When the gold–water interactions increase for lower loadings ($\varphi = 3.43$ % and $\varphi = 6.77$ %), the viscosity increases, while the rheology remains unchanged, as shown in Fig. 4.13. Therefore, the mechanism for the solidification effect, which is more likely at high loadings, is related to but differs from the mechanism for the solid-like absorbed water layer that alters the viscosity.

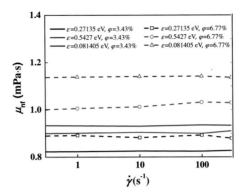

Fig. 4.13 Effects of gold nanoparticle–water molecule interactions on rheology for $\varphi = 3.43$ % and $\varphi = 6.77$ % nanoparticle loadings. Reprinted from Ref. [11], with kind permission from Springer Science+Business Media

4.4 Conclusions

The surface tension, viscosity, and rheology of gold–water nanofluids were examined using molecular dynamic simulations. The effects of the gold nanoparticle loadings and particle wettability were examined. The main conclusions are as follows:

1. The microscopic mechanism for the surface tension variation is related to the nanoparticle wettability. The repulsion of water molecules by surfactant-like nanoparticles increases the molecular spacing on the free surface which reduces the fluid surface tension. The attraction induced by non-surfactant nanoparticles reduces the molecular spacing on the free surface which increases the surface tension.
2. A solid-like absorbed water layer around the nanoparticles increases the equivalent nanoparticle radius and hinders the nanoparticle mobility within the base fluid which increases the nanofluid viscosity.
3. The nanofluid rheological behavior depends on the nanoparticle loading. For low loadings, the viscosity increases with increasing gold–water interaction forces, but remains unchanged with the shear rate, indicating Newtonian behavior. For high loadings, water molecule solidification is observed between neighboring nanoparticles due to the strong particle–particle interactions. These solidification structures are then disrupted for shear rates exceeding a critical value, which leads to shear-thinning non-Newtonian rheological behavior.

References

1. Choi SUS (1995) Enhancing thermal conductivity of fluids with nanoparticles, developments and application of non-newtonian flows. ASME, New York, FED 231/MD, vol 66, pp 99–105
2. Choi SUS (2009) Nanofluids: from vision to reality through research. J Heat Transfer 131:033106
3. Cheng LS, Cao DP (2011) Designing a thermo-switchable channel for nanofluidic controllable transportation. ACS Nano 5:1102–1108
4. Michaelides EE (2013) Transport properties of nanofluids. A critical review. J Non-Equilib Thermodyn 38:1–79
5. Chakraborty S, Padhy S (2008) Anomalous electrical conductivity of nanoscale colloidal suspensions. ACS Nano 2:2029–2036
6. Trisaksri V, Wongwises S (2007) Critical review of heat transfer characteristics of nanofluids. Renew Sustain Energy Rev 11:512–523
7. Branson BT, Beauchamp PS, Beam JC et al (2013) Nanodiamond nanofluids for enhanced thermal conductivity. ACS Nano 7:3183–3189
8. Wu S, Nikolov A, Wasan D (2013) Cleansing dynamics of oily soil using nanofluids. J Colloid Interface Sci 396:293–306
9. Murshed SMS, Leoong KC, Yang C (2008) Thermophysical and electrokinetic properties of nanofluids-A critical review. Appl Thermal Eng 28:2109–2125
10. Li YJ, Zhou JE, Tung S et al (2009) A review on development of nanofluid preparation and characterization. Powder Tech 196:89–101
11. Lu G, Duan YY, Wang XD (2014) Surface tension, viscosity, and rheology of water-based nanofluids: a microscopic interpretation on the molecular level. J Nanopart Res 16:2564
12. Zhu D, Wu S, Wang N (2010) Thermal physics and critical heat flux characteristics of Al_2O_3-H_2O nanofluids. Heat Transfer Eng 31:1213–1219
13. Moosavi M, Goharshadi EK, Youssefi A (2010) Fabrication, characterization, and measurement of some physicochemical properties of ZnO nanofluids. Int J Heat Mass Transfer 31:599–605
14. Tanvir S, Li Q (2012) Surface tension of nanofluid-type fuels containing suspended nanomaterials. Nanoscale Res Lett 7:226–236
15. Kumar R, Milanova D (2009) Effect of surface tension on nanotube nanofluids. Appl Phys Lett 94:073107
16. Chen RH, Phuoc TX, Martello D (2011) Surface tension of evaporating nanofluid droplets. Int J Heat Mass Transfer 54:2459–2466
17. Das SK, Putra N, Reotzel W (2003) Pool boiling characteristics of nano-fluids. Int J Heat Mass Transfer 46:851–862
18. Vafaei S, Purkayastha A, Jain A (2009) The effect of nanoparticles on the liquid-gas surface tension of Bi_2Te_3 nanofluids. Nanotechnology 20:185702
19. Murshed SM, Tan SH, Nguyen NT (2008) Temperature dependence of interfacial properties and viscosity of nanofluids for droplet-based microfluidics. J Phys D: Appl Phys 41:085502
20. Radiom M, Yang C, Chan WK (2010) Characterization of surface tension and contact angle of nanofluids. Proc SPIE 7522:75221D
21. Liu Y, Kai D (2012) Investigations of surface tension of binary nanofluids. Adv Mater Res 347–353:786–790
22. Einstein A (1956) Investigations on the theory of Brownian movement. Dover Publications, Inc, New York
23. Krieger IM, Dougherty TJ (1959) A mechanism for non-Newtonian flow in suspension of rigid spheres. Trans Soc Rheol 3:137–152
24. Nielsen LE (1970) Generalized equation for the elastic moduli of composite materials. J Appl Phys 41:4626
25. Batchelor GK (1977) The effect of Brownian motion on the bulk stress in a suspension of spherical particles. J Fluid Mech 83:97–117

26. Chandrasekar M, Suresh S, Chandra BA (2010) Experimental investigations and theoretical determination of thermal conductivity and viscosity of Al_2O_3/water nanofluids. Exp Thermal Fluid Sci 34:210–216

27. Nguyen CT, Desgranges F, Galanis N et al (2008) Viscosity data for Al_2O_3-water nanofluid-hysteresis: is heat transfer enhancement using nanofluids reliable. Int J Therm Sci 47:103–111

28. Lee JH, Hwang KS, Janga S et al (2008) Effective viscosities and thermal conductivities of aqueous nanofluids containing low volume concentrations of Al_2O_3 nanoparticles. Int J Heat Mass Transfer 51:2651–2656

29. Eastman JA, Phillpot SR, Choi SUS et al (2004) Thermal transport in nanofluids. Annual Rev Mater Res 34:219–246

30. Keblinski P, Eastman JA, Cahill DG (2005) Nanofluids for thermal transport. Mater Today 8:36–44

31. Mahbubul IM, Saidur R, Amalina MA (2012) Latest developments on the viscosity of nanofluids. Int J Heat Mass Transfer 55:874–885

32. Masoumi N, Sohrabi N, Behzadmehr A (2009) A new model for calculating the effective viscosity of nanofluids. J Phys D Appl Phys 42:055501

33. Hosseini SM, Moghadassi AR, Henneke DE (2010) A new dimensionless group model for determining the viscosity of nanofluids. J Therm Anal Calorim 100:873–877

34. Chen HS, Ding YL, Tan CQ (2007) Rheological behaviour of nanofluids. New J Phys 9:367

35. Prasher R, Song D, Wang J et al (2006) Measurements of nanofluid viscosity and its implications for thermal applications. Appl Phys Lett 89:133108

36. Chen HS, Ding YL, He YR et al (2007) Rheological behaviour of ethylene glycol based titaniananofluids. Chem Phys Lett 444:333–337

37. Susan-Resiga D, Socoliuc V, Boros T et al (2012) The influence of particle clustering on the rheological properties of highly concentrated magnetic nanofluids. J Colloid Interface Sci 373:110–115

38. Wang XW, Xu XF, Choi SUS (1999) Thermal conductivity of nanoparticle-fluid mixture. J Thermo Heat Transfer 13:474–480

39. He Y, Jin Y, Chen HS et al (2007) Heat transfer and flow behaviour of aqueous suspensions of TiO_2 nanoparticles (nanofluids) flowing upward through a vertical pipe. Int J Heat Mass Transfer 50:2272–2281

40. Murshed SMS, Leong KC, Yang C (2008) Investigations of thermal conductivity and viscosity of nanofluids. Int J Therm Sci 47:560–568

41. Chen HS, Ding YL, Lapkin A (2009) Rheological behaviour of nanofluids containing tube rod-like nanoparticles. Powder Tech 194:132–141

42. Ding Y, Alias H, Wen D, Williams RA et al (2006) Heat transfer of aqueous suspensions of carbon nanotubes (CNT nanofluids). Int J Heat Mass Transfer 49:240–250

43. Kole M, Dey TK (2011) Effect of aggregation on the viscosity of copper oxide–gear oil nanofluids. Int J Therm Sci 50:1741–1747

44. Yu W, Xie H, Li Y et al (2011) Experimental investigation on thermal conductivity and viscosity of aluminum nitride nanofluid. Particuology 9:187–191

45. Yu W, Xie H, Chen L et al (2009) Investigation of thermal conductivity and viscosity of ethylene glycol based ZnO nanofluids. Thermochim Acta 491:92–96

46. Kim S, Kim C, Lee WH et al (2011) Rheological properties of alumina nanofluids and their implication to the heat transfer enhancement mechanism. J Appl Phys 110:34316

47. Abareshi M, Sajjadi SH, Zebarjad SM et al (2011) Fabrication, characterization, and measurement of viscosity of α-Fe_2O_3-glycerol nanofluids. J Mol Liq 163:27–32

48. Shi B, Sinha S, Dhir VK (2006) Molecular dynamics simulation of the density and surface tension of water by particle-particle particle-mesh method. J Chem Phys 124:204715

49. Mountain RD (2009) An internally consistent method for the molecular dynamics simulation of the surface tension: application to some tip4p-type models of water. J Phys Chem B 113:482–486

50. Zhu RZ, Yang H (2011) A new method for the determination of surface tension from molecular dynamics simulations applied to liquid droplets. Chinese Phys B 20:016801
51. Rutkevych PP, Ramanarayan H, Wu DT (2010) Optimizing the computational efficiency of surface tension estimates in molecular dynamics simulations. Comp Mater Sci 49:s95–s98
52. Hou HY, Chen GL, Chen G (2009) A molecular dynamics simulation on surface tension of liquid Ni and Cu. Comp MaterSci 46:516–519
53. Sunda AP, Venkatnathan A (2013) Parametric dependence on shear viscosity of SPC/E water from equilibrium and non-equilbrium molecular dynamics simulations. Mol Simul 39:728–733
54. Medina JS, Prosmiti R, Villarreal P (2011) Molecular dynamics simulations of rigid and flexible water models: temperature dependence of viscosities. Chem Phys 388:9–18
55. Thomas JC, Rowley RL (2011) Transient molecular dynamics simulations of liquid viscosity for nonpolar and polar fluids. J Chem Phys 134:024526
56. Li X, Hede T, Tu Y (2011) Glycine in aerosol water droplets: a critical assessment of Köhler theory by predicting surface tension from molecular dynamics simulations. Atmos Chem Phys 11:519–527
57. D'Auria R, Tobias DJ (2009) On the relation between surface tension and ion adsorption at the air-water interface: a molecular dynamics simulation study. J Phys Chem A 113:7286–7293
58. Ge S, Zhang XX, Chen M (2011) Viscosity of NaCl aqueous solution under supercritical conditions: a molecular dynamics simulation. J Chem Eng Data 56:1299–1304
59. Chen T, Chidambaram M, Liu ZP et al (2010) Viscosities of the mixtures of 1-ethyl-3-methylimidazolium chloride with water, acetonitrile and glucose: a molecular dynamics simulation and experimental study. J Phys Chem B 114:5790–5794
60. Kumar P, Varanasi SR, Yashonath S (2013) Relation between the diffusivity, viscosity, and ionic radius of LiCl in water, methanol, and ethylene glycol: a molecular dynamics simulation. J Phys Chem B 117:8196–8208
61. Hasan B, Keblinski P, Khodadadi JM (2013) A proof for insignificant effect of Brownian motion-induced micro-convection on thermal conductivity of nanofluids by utilizing molecular dynamics simulations. J Appl Phys 113:084302
62. Mohebbi A (2012) Prediction of specific heat and thermal conductivity of nanofluids by a combined equilibrium and non-equilibrium molecular dynamics simulation. J Mol Liq 175:51–58
63. Cui WZ, Bai ML, Lv JZ (2011) On the influencing factors and strengthening mechanism for thermal conductivity of nanofluids by molecular dynamics simulation. Ind Eng Chem Res 50:13568
64. Li L, Zhang YW, Ma HB et al (2010) Molecular dynamics simulation of effect of liquid layering around the nanoparticle on the enhanced thermal conductivity of nanofluids. J Nanopart Res 12:811–821
65. Teng KL, Hsiao PY, Hung SW et al (2008) Enhanced thermal conductivity of nanofluids diagnosis by molecular dynamics simulations. J Nanosci Nanotech 8:3710–3718
66. Li L, Zhang YW, Ma HB et al (2008) An investigation of molecular layering at the liquid-solid interface in nanofluids by molecular dynamics simulation. Phys Lett A 372:4541–4544
67. Sarkara S, Selvam SP (2007) Molecular dynamics simulation of effective thermal conductivity and study of enhanced thermal transport mechanism in nanofluids. J Appl Phys 102:074302
68. Horn HW, Swope WC, Pitera JW et al (2004) Development of an improved four-site water model for biomolecular simulations: TIP4P-Ew. J Chem Phys 120:9665–9677
69. Ryckaert JP, Ciccotti G, Berendsen HJC (1977) Numerical integration of the cartesian equations of motion of a system with constraints: molecular dynamics of n-alkanes. J Comput Phys 23:327
70. Daw MS, Foiles SM, Baskes MI (1993) The embedded atom method: a review of theory and applications. Mater Sci Rep 9:251–310
71. Schravendijk P, van der Vegt N, Site LD et al (2005) Dual-scale modeling of benzene adsorption onto Ni(111) and Au(111) surfaces in explicit water. Chem Phys Chem 6:1866–1871

72. Plimpton S (1995) Fast parallel algorithms for short-range molecular dynamics. J Comput Phys 117:1–19
73. Ismail AE, Grest GS, Stevens MJ (2006) Capillary waves at the liquid-vapor interface and surface tension of water models. J Chem Phys 125:014702
74. Muller-Plathe F, Bordat P (2002) International summer school on novel methods in soft matter simulations. Helsinki Finland
75. Gittens GJ (1969) Variation of surface tension of water with temperature. J Colloid Interface Sci 30:406–412
76. Hilsenrath J, Klein M, Woolley HW (1955) Tables of thermal properties of gases. National Bureau of Standards Circular, Washington D.C
77. Einstein A (1905) Über die von der molecular kinetischen Theorie der Wärmege forderte Bewegung von in ruhenden Flüssig keiten suspendierten Teilchen. Ann Phys 322:549–560
78. Carré A, Woehl P (2006) Spreading of silicone oils on glass in two geometries. Langmuir 22:134–139

Chapter 5
Mesoscopic Studies of Nanofluid Dynamic Wetting: From Nanoscale to Macroscale

Abstract In this chapter, a lattice Boltzmann method with some simple but effective treatments with consideration of nanofluid surface tension and rheology modification, as well as the nanoparticle sedimentations, is conducted to investigate the effects of nanoparticle kinetics at the nanoscale (10^{-9} m) on the dynamic wetting behaviors occurring at the macroscopic scale (10^{-3} m). The study provides multi-scale understanding and guidelines to tune the nanofluid dynamic wetting behaviors.

5.1 Introduction

The mechanism of dynamic wetting by nanofluids is still unclear due to the complicated dynamic wetting behaviors, as well as the limitations of nanoscale experimental techniques and fundamental theories. In one aspect, the contact line motion is driven by the viscous force and surface tension force, which relate to the viscosity and surface tension of the spreading fluid. The nanoparticle motion in nanofluids significantly changes the surface tension, viscosity, and rheology of the base fluids [1, 2]. Therefore, the modification of the surface tension and the rheology by adding nanoparticles strongly affects the nanofluid dynamic wetting, which has been discussed as the bulk dissipation in Chap. 4. In other aspect, the suspensions of millions of nanoparticle induce a large amount of solid–liquid surface energy in nanofluids. To make the system steady, it is natural for nanoparticles to aggregate or self-assemble in the nanofluids to reach the lowest free energy state. The nanoparticle self-assembly near the contact line region strongly affects the contact line motion, leading to the nanofluid "super-spreading" behaviors [3], which has been discussed as the local dissipation in Chap. 3. The self-assembly of nanoparticles in the vicinity of the contact line region induces an additional disjoining pressure, braking the force balance at the contact line, and hence facilitating the motion of contact line [3–6]. The nanofluid "super-spreading" was widely used to explain the extraordinary nanofluid evaporation [7, 8] and boiling

[9–11]. Do these three effects, surface tension modification, rheology modification, and self-assembly, affect the dynamic wetting behavior individually or jointly? The answers to this question will reveal the mechanism of nanofluid dynamic wetting or provide the guideline to tune the nanofluid wetting kinetics. However, the study of the dynamic wetting by nanofluids is of great challenge since the wetting behavior crosses several length and timescales. Both the bulk and the local dissipation due to nanoparticle motions and self-assembly occur in the nanoscale. However, the spreading droplets usually have several millimeters in diameters. The difference between these two length scales is as large as several million times, from 10^{-9} to 10^{-3} m. Multiscale experimental techniques and fundamental theories are still insufficient to provide good understanding for these multiscale problems. The Lattice Boltzmann method (LBM) has been regarded as a very promising method to simulate the multiscale problems. The LBM is based on mesoscopic kinetic equations (the Boltzmann equation). The Navier–Stokes equations that describe the macroscopic flow problems can be derived from the Boltzmann equation using the Chapman–Enskog multiscale expansion. Due to the intrinsic microscopic kinetics, the LBM was widely used to investigate the flow and heat transfer of nanofluids [12–24] or pure fluid dynamic wetting [25–29]. However, there are few studies on the nanofluid dynamic wetting, especially the multiscale mechanisms from nanoparticle motions to macroscopic dynamic wetting.

In this chapter, the effects of nanoscale dissipations due to nanoparticle kinetics on the macroscopic dynamic wetting were studied from the viewpoint of mesoscopic simulations. The nanoscale dissipations contain the bulk dissipation, due to the nanoparticle motion in the bulk liquid, and the local dissipation due to the self-assembly of nanoparticles in the vicinity of the contact line regions. The roles of nanoparticle bulk and local dissipations on the macroscopic dynamic wetting were investigated by examining the individual or coupled effects of nanofluid surface tension, rheological properties, and nanoparticle self-assembly modes. The study provides multiscale understanding and tunable methods of the nanofluid dynamic wetting.

5.2 Simulation Models

5.2.1 Geometry Model

The simulated schematic is shown in Fig. 5.1. A 2D lattice grid was used to simulate a droplet spreading on an ideal and smooth solid surface. Three grid sizes, (I) 500×100 lu^2 (lattice unit, lu), (II) 1000×200 lu^2, and (III) 2000×400 lu^2 were used to test the grid independence, as shown in Fig. 5.1b. The moderate grid of 1000×200 lu^2 was used by considering both accuracy and efficiency. The droplet initial radius is $R_0 = 50$ lu. The left side and the right side are the period boundary conditions. The bounce-back condition is applied at the top side.

Fig. 5.1 Simulation geometry and lattice grid test. **a** Simulation geometry. **b** Lattice grid test

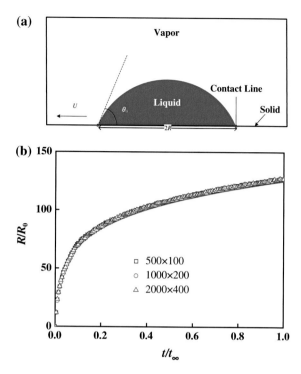

5.2.2 Dynamic Wetting with Triple-Phase Contact Line Motions

The droplet dynamic wetting is a replacement process of the liquid over the vapor on the solid surface, including interactions between the solid, the liquid, and the vapor surfaces. The multiphase LBM was used to simulate the replacement processes between the liquid and the vapor, including all the interactions between the solid–vapor, solid–liquid, vapor–vapor, liquid–liquid, and liquid–vapor.

A D2Q9 LBM model with Bhatnagar–Gross–Krook (BGK) collision operator was used to simulate the fluid flow [30]. To simulate multiphase fluids (liquid and vapor), the long-range interactions for liquid–liquid, vapor–vapor, and liquid–vapor are described using the Shan-Chen model [31–33]:

$$F(\mathbf{x}, t) = -G\psi(\mathbf{x}, t) \sum_{\alpha=1}^{8} w_\alpha \psi(\mathbf{x} + \mathbf{e}_\alpha \Delta t, t)\mathbf{e}_\alpha, \tag{5.1}$$

where G is the interaction strength and ψ is the interaction potential expressed as [34]:

$$\psi(\rho) = \psi_0 \exp\left(\frac{-\rho_0}{\rho}\right), \tag{5.2}$$

where ψ_0 and ρ_0 are constants, usually with the values of 4 and 200 [35].

The dynamic wetting of fluids on the solid surfaces is governed by an adhesive interaction force between the fluid (liquid or vapor) and the solid surface [36]:

$$F_{ads}(\mathbf{x}, t) = -G_{ads}\psi(\mathbf{x}, t) \sum_{\alpha=1}^{8} w_\alpha s(\mathbf{x} + \mathbf{e}_\alpha \Delta t, t) \mathbf{e}_\alpha, \tag{5.3}$$

where G_{ads} is the adsorption coefficient, which is 327 in this study. s is a switch parameter with the value of 1, if the neighbor lattice of the fluid lattice is a solid boundary and 0 for a neighbor fluid lattice.

5.2.3   Nanofluid Modeling

Figure 5.2 shows the two nanoparticle dissipation mechanisms due to the nanoparticle motion and the self-assembly: bulk dissipation and local dissipation. In the bulk dissipation, nanoparticles distribute homogeneously in the bulk liquids. Therefore, it is reasonable to assume that the nanofluids are homogeneous single-phase. The nanoparticles were treated as "fluid" lattices. The particle–particle and particle–fluid interactions are still cohesive forces, which can be calculated with Eq. (5.1) with different G. In the simulations, G switches to the particle–fluid/particle–particle value if the lattice or the neighbor lattice is occupied by a nanoparticle lattice. The nanoparticles only modify the surface tension and rheology of the base fluids if the bulk dissipation occurs only. For the local dissipation, the nanoparticle lattices deposit at a given deposition rate to the bottom or to the vicinity of the contact line region during the droplet spreading, changing the thickness of liquid film and then resulting the additional disjoining pressures, which brake the resultant forces and then facilitate the motion of contact line.

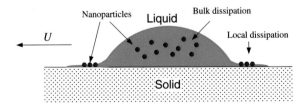

Fig. 5.2 Schematic of the two nanoparticle dissipations of nanofluid dynamic wetting

5.2.3.1 Bulk Dissipation Due to the Surface Tension and Rheology Modification

The nanoparticles significantly modify the surface tension of nanofluids, which strongly affects the dynamic wetting process [37]. The nanoparticles can increase or reduce the surface tension of base fluids, which depends on the interactions between nanoparticles and solvent molecules. The molecular dynamic simulations show that the nanoparticle wettability is responsible for the surface tension modifications [38]. Hydrophobic nanoparticles always tend to stay on the free surface so they behave like a surfactant to reduce the surface tension. Hydrophilic nanoparticles immerge into the bulk fluid which increases the surface tension of the nanofluids. Thus, at mesoscopic scale level, the surface tension modifications of nanofluids due to the nanoparticle dissipations were simulated by changing the fluid–fluid "particle" interaction strength G in Eq. (5.1). Two typical nanoparticles, hydrophobic and hydrophilic, were calculated with different G. The pure fluid was calculated as the baseline with $G = -130$, for which the surface tension is 15.8 μ ls^{-2}. For a 2D nanofluid drop with an initial radius of 50 lattices and 6 % nanoparticle volume fraction, the nanoparticles occupied 471 lattices (7854 lattices in total for the nanofluid drop with $R_0 = 50$ lu). By switching the particle–particle/particle–fluid interactions for these 471 lattices or their neighbor lattices to $G = -120$, the nanofluids with adding hydrophobic nanoparticles can be simulated, with a surface tension of 14.3 μ ls^{-2}. The nanofluid with hydrophilic nanoparticles, simulated with the particle–particle/particle–fluid interaction parameter of $G = -140$, has a surface tension of 17.6 μ ls^{-2}. The properties of the three types of fluids are listed in Table 5.1.

The adding of nanoparticles also modifies the rheological behavior of the base fluid, especially for the high nanoparticle loadings [1]. The rheology modification of nanofluids was simulated in the LBM by changing the relaxation time τ_f based on the local shear rates at every iterative step. The power-law rheological behavior was described by

$$\mu = \kappa\dot{\gamma}^{(n-1)}, \tag{5.4}$$

where κ is the viscosity coefficient, n is the rheological index, and $\dot{\gamma}$ is the local shear rate, which is calculated by the local shear rate tension $\mathbf{d}$,

$$\dot{\gamma}_{x,y} = \sqrt{2\mathbf{d}{:}\mathbf{d}}, \tag{5.5}$$

Table 5.1 Surface tension modification by nanoparticles with various wettabilities

Nanoparticle wettability	G	ρ_l	ρ_v	p_{sat}	σ
Hydrophobic	−120	524.37	85.7	25.56	14.3
Pure fluids	−130	611.27	76.16	23.57	15.8
Hydrophilic	−140	696.11	69.06	21.88	17.6

Fig. 5.3 Four shear-thinning
nanofluids

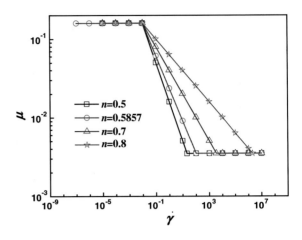

Table 5.2 Four
shear-thinning nanofluids
($\mu_0 = 0.16$, $\mu_\infty = 0.0035$)

n	Shear rate ranges	Parameters
$n = 0.5$	$10^{-2} < \dot{\gamma} < 21$	$k = 0.016$
$n = 0.5857$	$10^{-2} < \dot{\gamma} < 10^2$	$k = 0.0236$
$n = 0.7$	$10^{-2} < \dot{\gamma} < 3.4 \times 10^3$	$k = 0.04019$
$n = 0.8$	$10^{-2} < \dot{\gamma} < 2 \times 10^6$	$k = 0.06369$

where $\mathbf{d} = \frac{1}{2}(\nabla \mathbf{u} + \nabla \mathbf{u}')$.

The shear-thinning non-Newtonian behaviors were observed in most nanofluids
[7]. The truncated power law (TPL) was used to describe the relationships of
viscosities and shear rates. To examine the effects of rheology on the dynamic
wetting, four shear-thinning nanofluids, with the equal viscosity boundaries
($\mu_0 = 0.16$, $\mu_\infty = 0.0035$) but different rheological indexes, are shown in Fig. 5.3.
The TPL parameters of the four shear-thinning nanofluids are listed in Table 5.2.

It should be noted that the nanoparticle self-assembly is not considered when the
bulk dissipations are discussed.

5.2.3.2 Structural Disjoining Pressure Due to Nanoparticle Self-assembly

Figure 5.4 shows two nanoparticle self-assembly modes for both the complete and
the partial wetting. Nanoparticles likely deposit at the bottom (global deposition) or
near the contact line (local deposition). The bottom and the contact line region
depositions were regarded as two enhancement mechanisms of nanofluid dynamic
wetting [7]. The structural disjoining pressure was used to explain the effects of
nanoparticle deposition on the wetting kinetics [4]. A linear deposition rate with
de = 0.007 lu^2/lt is given in Fig. 5.5, with which 141 fluid lattices transfer into solid

Fig. 5.4 Various
nanoparticle self-assembly
modes: **a** bottom deposition
for the complete wetting;
b contact line region
deposition for the complete
wetting; **c** bottom deposition
for the partial wetting;
d contact line region
deposition for the partial
wetting

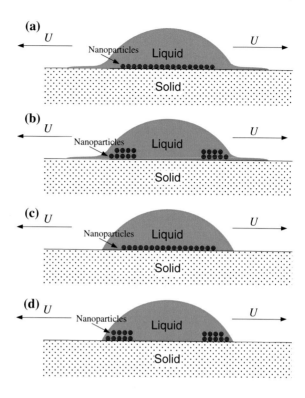

Fig. 5.5 Nanoparticle
deposition rate

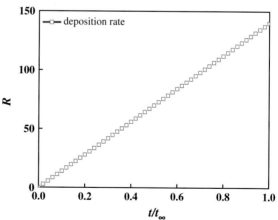

lattices on the lowest and the second lowest lattice layer, corresponding to 30 %
nanoparticle deposit at $t = 20{,}000$ lt. As a nanoparticle lattice transfers into a solid
lattice at the solid surface, a random particle lattice in the bulk liquid droplet
changes into fluid lattice, correspondingly. For the bottom deposition cases, the
nanoparticles are assumed to deposit from the center of drop bottom, and then

radiate outwards to the contact line region. If the lowest lattice layer was completely occupied by the deposited nanoparticles, the second lowest lattice layer transfers to the solid lattice from center to the droplet margin according to the given nanoparticle deposition rate. For the contact line deposition cases, the fluid lattices, with 1 lattice away from the contact line lattices, transfer to the solid lattices if they are occupied by the deposited nanoparticles. As the contact line move forwards, the second deposited nanoparticle appears at the location with 1 lattice away from the contact line. If more than two nanoparticles deposit at one time step, one nanoparticle deposits on the lowest layer, and the other deposits on the upper layer. It is noted that the bulk dissipation is neglected in the cases of nanoparticle sedimentation or self-assembly.

In nanoscale, the strong interactions between the substrate and the liquid molecule reduce the liquid pressure within the thin film and result in the disjoining pressure [38].

The disjoining pressure is defined as follows [39]:

$$\Pi = -\frac{A_{SL}}{6\pi\delta^3} \tag{5.6}$$

where A_{SL} is the Hamaker constant, and δ is the thickness of liquid film. The disjoining pressure decreases with increasing the liquid film thickness. As discussed in Chap. 3, nanoparticles self-assemble at the bottom or near the contact line region reducing the thickness of the thin film, leading to an additional disjoining pressure called structural disjoining pressure, which facilitates the motion of the contact line. The disjoining pressure can also be calculated using the pressure equilibrium on the liquid–vapor interface [40],

$$\Pi = \rho k_B T \ln\left(\frac{p_v}{p_{sat}}\right) \tag{5.7}$$

where p_v is the vapor pressure, p_{sat} is the saturated pressure, k_B is the Boltzmann constant, T is the temperature, and ρ is the vapor density.

Figure 5.6 shows the condensation of vapor with different saturabilities within the capillary tube. The thickness of condensed liquid film increases with increasing vapor saturability. Figure 5.7 shows the comparison of the disjoining pressures calculated from Eq. (5.6) using the LBM-simulated film thickness with the results from Eq. (5.7) using the vapor saturability. The good agreement indicates that the LBM can predict the disjoining pressure which occurs in nanoscale. Figure 5.7 also indicates that the disjoining pressure only exists in the thin film with less than 2 lattices in thickness. The deposited nanoparticles reduce the thin film thickness and result in the structural disjoining pressure.

The individual effect of the surface tension, rheology, and the structural disjoining pressure was studied to examine the bulk and local dissipation effects due to nanoparticle motion on the nanofluid dynamic wetting.

Fig. 5.6 Condensation of
vapor within capillary tube:
a vapor saturability $\varphi = 5$ %;
b $\varphi = 10$ %; **c** $\varphi = 15$ %;
d $\varphi = 20$ %

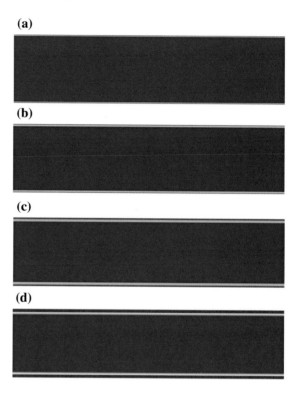

Fig. 5.7 Comparison of
disjoining pressure calculated
with simulated film thickness
(LBM simulation) and the
calculation with the given
vapor saturation

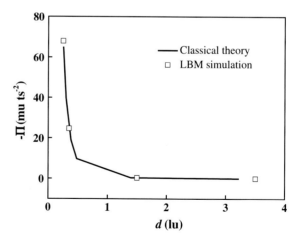

5.3   Results and Discussions

5.3.1   *Bulk Dissipation Due to the Surface Tension Modification*

Figure 5.8 shows the pure fluid and nanofluid dynamic wetting at $t = 18,000$ lt (lattice time). The nanofluids with hydrophobic nanoparticle, as shown in Fig. 5.8a, prefer to spread completely, while the pure fluids (Fig. 5.8b) and the nanofluids with hydrophilic nanoparticles (Fig. 5.8c) show the partial wetting behavior. The equilibrium contact angle of the nanofluids with hydrophilic nanoparticles is larger than that of pure fluids. The precursor layer was only observed in the nanofluids with hydrophobic nanoparticles. The nanoparticle wettability is responsible for the nanofluid surface tension. The surface tension of base fluids decreases with adding hydrophobic nanoparticles, but increases with adding hydrophilic nanoparticles [38]. According to the contact line motion driven force, $F = \sigma_{SL} - \sigma_{SV} - \sigma_{LV}\cos\theta$, the spreading increases with decreasing the liquid–vapor surface tension.

Figure 5.9a shows the non-dimensional spreading radius variation with spreading time, in which $R_0 = 50$ lu and $t_\infty = 20,000$ lt. Both the pure fluid and the nanofluid with hydrophilic nanoparticles have reached the equilibrium stages. The equilibrium non-dimensional radius of the pure fluid droplet was $R/R_0 = 1.653$ after $t/t_\infty = 0.530$, while $R/R_0 = 1.384$ for the nanofluid drop with hydrophilic nanoparticles after $t/t_\infty = 0.445$. The nanofluid drop with hydrophobic nanoparticles has a spreading radius of 2.529 and keep spreading after $t_\infty = 20,000$ lt. The logarithmic results in Fig. 5.9b indicate the two stages of the droplet spreading, the initial fast spreading stage and the slow spreading stage. The two stage partial wetting was also reported in the Ref. [41]. The slow spreading stages occupy most of the spreading time, for insistence, 92 % for the hydrophobic nanoparticle fluids,

Fig. 5.8 Effects of nanoparticle wettability on the dynamic wetting of nanofluids ($t = 18,000$ lt): **a** hydrophobic nanoparticles; **b** pure fluids; **c** hydrophilic nanoparticles

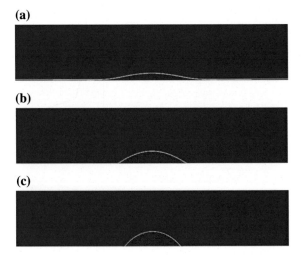

(a)

(b)

(c)

Fig. 5.9 Non-dimensional spreading radii versus spreading time for the three fluid droplets

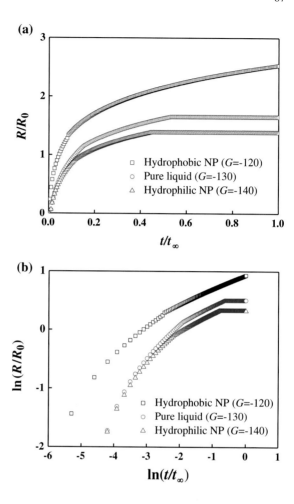

85 % for pure fluids, and 89 % for hydrophilic nanoparticle fluids. The spreading laws of the three fluid droplets are fitted as shown below:

$$
\begin{aligned}
R' &\sim t'^{0.145} & & (G = -120) \\
\begin{cases} R' \sim t'^{0.261} & (t' < 0.145) \\ R' \sim t'^{0.142} & (t' > 0.145) \end{cases} & & & (G = -130) \\
\begin{cases} R' \sim t'^{0.241} & (t' < 0.115) \\ R' \sim t'^{0.140} & (t' < 0.115) \end{cases} & & & (G = -140)
\end{aligned}
\tag{5.8}
$$

where the spreading exponent of the complete wetting (CW) ($G = -120$), $\alpha = 0.145$, is larger than 1/7, and the spreading exponent of 2D Newtonian fluid droplets is predicted by the hydrodynamic model. The results indicate that the additional hydrophobic nanoparticles, which acting like surfactants, greatly facilitate the dynamic wetting, while hydrophilic nanoparticles inhibit the droplet spreading. The results also indicate that the dynamic wetting by nanofluids with hydrophobic nanoparticles deviate from Newtonian spreading behaviors. The effects of rheology on the dynamic wetting should be examined.

5.3.2 Bulk Dissipation Due to the Rheology Modification

To examine the effects of rheology on the nanofluid dynamic wetting, two additional types of fluids were simulated. One fluid is created with $G = -130$ for all particle–particle/particle–fluid/fluid–fluid interactions and the rheological exponent of $n = 0.8$, corresponding to a shear-thinning non-Newtonian fluid without nanoparticles. The other fluid is simulated with $G = -120$ and $n = 0.8$, corresponding to a shear-thinning non-Newtonian fluid with hydrophobic nanoparticles. Figure 5.10 shows the effects of surface tension and rheology modification on the dynamic wetting of nanofluids. The results indicate that the dynamic wetting by nanofluids is dominated by both the surface tension and the rheological properties. The dynamic wetting capacity is underestimated if the rheology modification is neglected, as compared $G = -120$, $n = 0.8$ with $G = -120$, $n = 1$. However, if the surface tension modification is neglected, the nanofluids exhibit the partial wetting behavior, as shown in the comparison between $G = -120$, $n = 0.8$ and $G = -130$, $n = 0.8$. The modification of surface tension is more related to the wettability capacity which can be characterized by the spreading coefficient, $S = \sigma_{SL} - \sigma_{SV} - \sigma_{LV}$. The fluids exhibit CW behaviors if $S > 0$ while partial wetting for $S < 0$. Although the rheology is modified for the fluids with $G = -130$ and $n = 0.8$, the fluids still exhibit partial wetting behaviors. Therefore, both the modifications of surface tension and rheology contribute to the dynamic wetting by nanofluids.

The effects of rheology on the dynamic wetting are further examined by changing the rheological index. The adding nanoparticles are hydrophobic

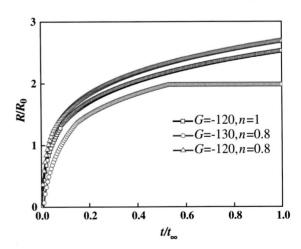

Fig. 5.10 Effects of surface tension and rheology on the dynamic wetting of nanofluids

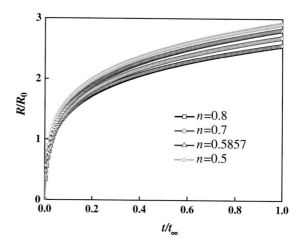

Fig. 5.11 Effects of rheological indexes on the dynamic wetting of nanofluids ($G = -120$)

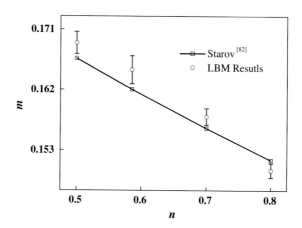

Fig. 5.12 Comparison of LBM results and Starov's model for the spreading exponents versus rheological index

($G = -120$). As shown in Fig. 5.11, the dynamic wetting capability increases with decreasing the rheological indexes. The dynamic wetting is enhanced due to the weaker viscous dissipation. Figure 5.12 shows the relation of the spreading exponents and the rheological indexes. The LBM results agree with the Starov's non-Newtonian dynamic wetting model [42], in which $m = n/(5 + 2n)$ for a 2D droplet spreading.

5.3.3 Local Dissipation Due to Structural Disjoining Pressure

Figure 5.13 shows the effects of nanoparticle self-assembly on the dynamic wetting of nanofluids at $t = 20,000$ lt, in which CW is the CW (by adding hydrophobic

Fig. 5.13 Effects of nanoparticle self-assembly on the dynamic wetting process

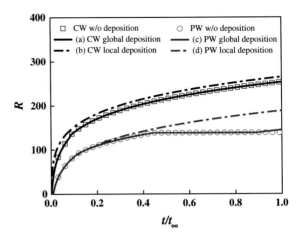

nanoparticles), PW is the partial wetting (by adding hydrophilic nanoparticles), and w/o presents without deposition. It should be noted that the rheology modification is not considered here because the rheology does not change the wetting behaviors qualitatively as discussed in Sect. 5.3.2. For the CW drop, the globally deposited nanoparticles do not affect the thin film thickness in the vicinity of the contact line region during the nanofluid droplet spreading, even for the 30 % deposition. Therefore, the deposition has few effects on the dynamic wetting process. However, the local deposition in the vicinity of the contact line region reduces the thin film thickness, leading to an additional disjoining pressure. Therefore, the local deposition enhances the dynamic wetting process from the beginning of wetting process. For the partial wetting droplet, the globally deposited nanoparticles do not affect the wetting kinetics at the beginning. At the last stage ($t/t_\infty > 0.9$), when the deposited nanoparticles change the thin film thickness, the spreading is enhanced due to the nanoparticle deposition. For the local deposition, since the dynamic contact angle is large, the nanoparticle deposition has few effects on the dynamic wetting at the beginning stage. However, the deposition strongly affects the dynamic wetting at the last stage, since the deposition of nanoparticles reduces the thin film thickness for the small contact angle cases, leading to an additional structural disjoining pressure near the contact line region. Figure 5.14 shows the disjoining pressure in the vicinity of the contact line region for the four nanoparticle deposition modes. The disjoining pressure difference between the contact line and the bulk liquid region is responsible for the motion of contact line. For the CW, the nanoparticle depositions have few effects on the disjoining pressure. However, for the partial wetting, the depositions strongly affect the disjoining pressure in the vicinity of the contact line region. Therefore, the effects of nanoparticle deposition for the partial wetting are more significant than that of CW. For the CW, the disjoining pressure in the precursor layer balances the structural disjoining pressure due to the self-assembly of nanoparticles, hence the nanoparticle self-assembly has few effects on the dynamic wetting.

Fig. 5.14 Disjoining pressure in the vicinity of contact line region for the four nanoparticle deposition modes ($t = 20,000$ lt)

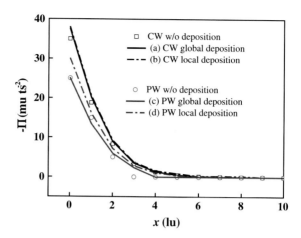

5.4 Conclusions

The multiphase LBM was used to study the nano-scale dissipation effects on the macroscale dynamic wetting process. The microscopic dissipation effects include the surface tension and the rheology modifications due to the nanoparticle motion in the bulk liquid, as well as the structural disjoining pressure due to the nanoparticle deposition near the contact line region.

1. The adding of hydrophobic nanoparticle facilitates the dynamic wetting while the hydrophilic nanoparticles deteriorate the dynamic wetting. The partial wetting process was divided into two stages. The spreading exponent of the slow stage consists with the hydrodynamic model prediction;
2. The shear-thinning non-Newtonian behavior due to the adding of nanoparticle enhances the dynamic wetting of nanofluids. The wetting capability increases with the decreasing rheological indexes. The relation of the spreading exponent and the rheological index agree with the Starov's model.
3. The nanoparticle global deposition has few effects on the dynamic wetting, while the local deposition strongly affects the dynamic wetting process. For the partial wetting drop, the structural disjoining pressure due to the self-assembly of nanoparticle in the vicinity of the contact angle region greatly facilitates the contact line motion.

References

1. Chen HS, DingYL TanCQ (2007) Rheological behaviour of nanofluids. New J Phys 9:367
2. Tanvir S, Li Q (2012) Surface tension of nanofluid-type fuels containing suspended nanomaterials. Nanoscale Res Lett 7:226–236
3. Wasan DT, Nikolov AD (2003) Spreading of nanofluids on solids. Nature 423:156–159

4. Kondiparty K, Nikolov AD, Wu S et al (2011) Wetting and spreading of nanofluids on solid surfaces driven by the structural disjoining pressure: statics analysis and experiments. Langmuir 27:3324–3335
5. Kondiparty K, Nikolov AD, Wasan DT et al (2012) Dynamic spreading of nanofluids on solids part I: Experimental. Langmuir 28:14618–14623
6. Liu KL, Kondiparty K, Nikolov AD et al (2012) Dynamic spreading of nanofluids on solids part II: modeling. Langmuir 28:16274–16284
7. Sefiane K, Bennacer R (2009) Nanofluids droplets evaporation kinetics and wetting dynamics on rough heated substrates. Adv Colloid Interface Sci 147–148:263–271
8. Moffat JR, Sefiane K, Shanahan MER (2009) Effect of TiO$_2$ nanoparticles on contact line stick-slip behavior of volatile drops. J Phys Chem B 113:8860–8866
9. Murshed SMS, Nieto de Castro CA, Lourenco MJV et al (2007) A review of boiling and convective heat transfer with nanofluids. Renew Sust Energy Rev 15:2342–2354
10. Wen DS (2008) Mechanisms of thermal nanofluids on enhanced critical heat flux (CHF). Int J Heat Mass Transf 51:4958–4965
11. Sefiane K (2006) On the role of structural disjoining pressure and contact line pinning in critical heat flux enhancement during boiling of nanofluids. Appl Phys Lett 89:044106
12. Fattahi E, Farhadi M, Sedighi K (2011) Lattice Boltzmann simulation of natural convection heat transfer in nanofluids. Int J Therm Sci 50:137–144
13. Lai FH, Yang YT (2011) Lattice Boltzmann simulation of natural convection heat transfer of Al$_2$O$_3$/water nanofluids in a square enclosure. Int J Therm Sci 50:1930–1941
14. Yang YT, Lai FH (2011) Numerical study of flow and heat transfer characteristics of alumina-water nanofluids in a microchannel using the Lattice Boltzmann method. Int Commun Heat Mass Transf 38:607–614
15. Nabavitabatabayi M, Shirani E, Rahimian MH (2011) Investigation of heat transfer enhancement in an enclosure filled with nanofluids using multiple relaxation time Lattice Boltzmann modeling. Int Commun Heat Mass Transf 38:128–138
16. Bararnia H, Hooman K, Ganji DD (2011) Natural convection in nanofluids-filled portioned cavity: the Lattice-Boltzmann method. Numer Heat Transf A 59:487–502
17. Kefayati GR, Hosseinizadeh SF, Gorji M (2012) Lattice Boltzmann simulation of natural convection in an open enclosure subjugated to water/copper nanofluid. Int J Therm Sci 52:91–101
18. Kefayati GR, Hosseinizadeh SF, Gorji M (2011) Lattice Boltzmann simulation of natural convection in tall enclosures using water/SiO$_2$ nanofluid. Int Commun Heat Mass Transf 38:798–805
19. He YR, Qi C, Hu YW (2011) Lattice Boltzmann simulation of alumina-water nanofluid in a square cavity. Nanoscale Res Lett 6:184
20. Nemati H, Farhadi M, Sedighi K (2010) Lattice Boltzmann simulation of nanofluid in lid-driven cavity. Int Commun Heat Mass Transf 37:1528–1534
21. Xuan YM, Yu K, Li Q (2005) Investigation on flow and heat transfer of nanofluids by the thermal Lattice Boltzmann model. Prog Comput Fluid Dyn 5:13–19
22. Xuan YM, Li Q, Yao ZP (2004) Application of Lattice Boltzmann scheme to nanofluids. Sci China Ser E 47:129–140
23. Xuan YM, Yao ZP (2005) Lattice Boltzmann model for nanofluids. Heat Mass Transf 3:199–205
24. Zhou LJ, Xuan YM, Li Q (2010) Multiscale simulation of flow and heat transfer of nanofluid with Lattice Boltzmann method. Int J Multiph Flow 36:364–374
25. Shih CH, Wu CL, Chang LC et al (2011) Lattice Boltzmann simulations of incompressible liquid-gas systems on partial wetting surfaces. Philos Trans R Soc Lond Series A 369:2510–2518
26. Joshi AS, Sun Y (2010) Wetting dynamics and particle deposition for an evaporating colloidal drop: a Lattice Boltzmann study. Phys Rev E 82:041401
27. Yu Y, Liu YL (2008) Lattice-Boltzmann models simulation of wetting modes on the surface with nanostructures. J Comput Theor Nanosci 5:1377–1380

28. Yan YY, Zu YQ (2007) A Lattice Boltzmann method for incompressible two-phase flows on partial wetting surface with large density ratio. J Comput Phys 227:763–775
29. Davies AR, Summers JL, Wilson MCT (2006) On a dynamic wetting model for the finite-density multiphase Lattice Boltzmann method. Int J Comput Fluid Dyn 20:415–425
30. Zou Q, Hou S, Chen S (1995) An improved incompressible Lattice Boltzmann model for time-independent flows. J Stat Phys 81:35–48
31. Shan XW, Chen HD (1993) Lattice Boltzmann model for simulating flows with multiple phases and components. Phys Rev E 47:1815–1819
32. Shan XW, Chen HD (1994) Simulation of nonideal gases and liquid-gas phase-transitions by the Lattice Boltzmann-equation. Phys Rev E 49:2941–2948
33. Shan XW, Doolen GD (1995) Multicomponent Lattice-Boltzmann model with interparticle interaction. J Stat Phys 81:379–393
34. He XY, Doolen GA (2002) Thermodynamic foundations of kinetic theory and Lattice Boltzmann models for multiphase flows. J Stat Phys 107:309–328
35. Sukop MC, Thorne DT Jr (2007) Lattice Boltzmann modeling. Springer, New York
36. Martys NS, Chen HD (1996) Simulation of multicomponent fluids in complex three-dimensional geometries by the Lattice Boltzmann method. Phys Rev E 53:743–750
37. Lu G, Hu H, Duan YY et al (2013) Wetting kinetics of water nano-droplet containing non-surfactant nanoparticles: a molecular dynamics study. Appl Phys Lett 103:253104
38. Lu G, Duan YY, Wang XD (2014) Surface tension, viscosity, and rheology of water-based nanofluids: a microscopic interpretation on the molecular level. J Nanopart Res 16:2564
39. Israelachviliv JN (2011) Intermolecular and surface forces, 3rd edn. Academic Press, Burlington
40. Carry VP, Wemhoff AP (2006) Disjoining pressure effects in ultra-thin liquid films in micropassages—comparison of thermodynamic theory with predictions of molecular dynamics simulations. ASME J Heat Transf 128:1276–1284
41. de Ruijter MJ, de Coninck J, Oshanin G (1999) Droplet spreading: partial wetting regime revisited. Langmuir 15:2209–2216
42. Starov VM, Tyatyushkin AN, Velarde MG et al (2003) Spreading of non-Newtonian liquids over solid substrates. J Colloid Interface Sci 257:284–290

Chapter 6
Nanofluid Dynamic Wetting with External Thermal Fields

Abstract In this chapter, the nanofluid dynamic wetting was studied with complex external conditions. The effects of substrate heating and intensive free surface evaporation on the wetting kinetics of nanofluids were simulated using molecular dynamics simulations. The effects of initial droplet temperature, substrate temperature, and wettability were examined. The microscopic mechanisms of nanoparticle self-assembly and nanofluid droplet evaporating-spreading behavior were revealed by tracing the particle motion and molecular mobility near the contact line region.

6.1 Introduction

In this chapter, we will focus on the application of nanofluid dynamic wetting with external conditions. Dynamic wetting of fluids is of great importance in various practical applications such as coating, inkjet printing, and liquid–vapor phase change [1–3]. The previous studies mainly focused on the dynamic wetting of simply fluids (Newtonian fluids or simple rheology non-Newtonian fluids) on ideal surfaces without any external fields. In practice, the dynamic wetting process always occurs with complex external fields, such as thermal, electric, magnetic, or chemical. These fields greatly affect the flow and mass transport behaviors during the fluid wetting, hence affecting the contact line motion. For example, many experimental studies had revealed that free surface evaporation greatly affected the contact line motion and the dynamic contact angle [4–9]. However, the mechanism of dynamic wetting with evaporating is still unclear. In this chapter, we considered a nanometer droplet spreading with heat transfer and phase change, which is of practical interest in many thermal engineering systems or devices. For example, during the inkjet printing, the nanoscale or submicroscale ink droplets impact and spread on the papers with extremely high flux conditions, and the evaporation occurs strongly during the droplet dynamic wetting process. The resent nanoprinting is regarded as a very promising technique, which extends the study of dynamic wetting into the nanoscale [10].

© Springer-Verlag Berlin Heidelberg 2016
G. Lu, *Dynamic Wetting by Nanofluids*,
Springer Theses, DOI 10.1007/978-3-662-48765-5_6

The spreading of non-evaporating droplets has been widely studied experimentally and theoretically using macroscopic hydrodynamic models [11–13] and numerically using the macroscopic computational fluid dynamics (CFD) models [14–16] mesoscopic lattice Boltzmann method (LBM) [17–19], and microscopic molecular dynamic (MD) simulations [20–23]. The using of CFD models to analyze dynamic wetting is still facing challenges due to the "contact line paradox" [24]. The lack of an adequate phase change model has hindered the usage of LBM models in simulating the evaporating problems. Both the dynamic wetting and the evaporating are easily implemented in MD simulations with few assumptions. Therefore, MD simulations are promising tools to simulate the dynamic wetting with evaporating. Many MD studies have focused on the droplet wetting kinetics in constant temperatures [17–20] or droplet evaporating on heated solid surfaces without spreading [25, 26]. Hwang et al. [27] studied the effects of temperature on droplet spreading using molecular dynamic simulations with a Lennard-Jones (LJ) liquid spreading on an LJ solid. However, no details on the spreading–evaporating mechanism were discussed in their study. The study of nanofluid dynamic wetting with evaporating is more of practical interest since the process usually takes place in the heating environments. The nanoparticle depositing due to the coupled effects of the spreading and evaporating of nanofluids is a fundamental process in many industrial applications, such as printing, coating, or material manufacturing. For nanofluid droplets, the additional nanoparticles induce more complex solid–liquid molecular interactions. Therefore, the dynamic wetting with evaporating by nanofluids is different from the base fluids, which has been observed experimentally. However, there is little microscopic understanding of the effects of evaporation on the dynamic wetting by nanofluids and there are no good theoretical spreading–evaporating models.

The present works study the wetting kinetics of nanofluid droplets with heating conditions using molecular dynamic simulations. The effects of the evaporation on the spreading are examined by analyzing the nanoparticle distribution, contact line mobility, and the molecule absorption–desorption behavior in the vicinity of the contact line region. The effects of substrate temperature, initial droplet temperature, and droplet size, as well as the substrate wettability, on the nanofluid droplet spreading–evaporating processes are described in this work. The objectives are to explore the mechanisms of nanofluid dynamic wetting with evaporating conditions, as well as the tuning method of dynamic wetting with external fields.

6.2 Simulation System

6.2.1 Models

Figure 6.1 shows the simulation model for the nanofluid droplet (d = 10 nm, l = 1.6 nm, 4500 water molecules, nanoparticle volume fraction is 6.77 %)

Fig. 6.1 Schematic of
simulation setups of nanofluid
droplets spreading with
evaporating

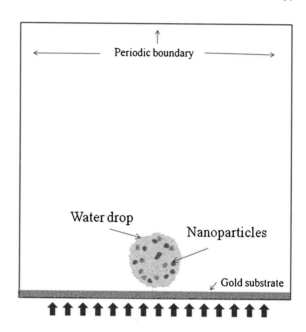

Table 6.1 Simulation cases

	Substrate temperature (K)	Droplet temperature (K)
Limited evaporation	300	300
Evaporation #1	333	300
Evaporation #2	363	300
Evaporation #3	363	333
Evaporation #4	363	363

Reprinted from Ref. [28], with kind permission from ASME

spreading on a (100) gold surface ($60 \times 1.6 \times 1.6$ nm^3, 9600 gold atoms). The
molecular dynamic simulation parameters were the same as given in Chap. 3. The
gold substrate structures were equilibrated at various temperatures from 300 to
363 K. The nanofluid droplets were initially equilibrated at 300, 333, and 363 K.
Then, the droplets touched and spread on the heated gold substrates in equilibrium
stages.

Two simulation systems were listed in Table 6.1 with constant initial droplet
temperature ($T_l = 300$ K), while the substrate was maintained at three temperatures
of 300, 333, and 363 K and a constant substrate temperature ($T_s = 363$ K), while the
water droplet was maintained at three temperatures of 300, 333, and 363 K.

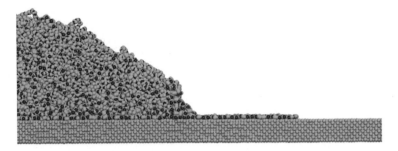

Fig. 6.2 Schematic of precursor layer for a spreading droplet. Reprinted from Ref. [28], with kind permission from ASME

6.2.2 Model Validation

The calculated results were validated by fitting the spreading exponent and solid–liquid friction coefficient for pure water droplets, as shown in Fig. 3.3 in Chap. 3. The dynamic wetting by nanofluid droplet is divided into two stages with different spreading power laws. The rapid spreading stage with R–$t^{0.466}$ is faster than that observed in experimental measurements or predicted by hydrodynamic models ($n = 1/10$ for the capillary governing regime and $n = 1/8$ for gravitational governing regime [2]). The difference mainly comes from the different statistical contact line approaches. Only apparent contact line can be observed in experimental measurements or hydrodynamic models. However, in MD simulations, the microscopic precursor layer with thickness of several angstroms can be observed as shown in Fig. 6.2, which obeys the spreading law of R–$t^{1/2}$ [1]. Therefore, the droplet spreads faster due to the motion of the precursor layer, leading to a larger spreading exponent of 0.466. The slower spreading stage, which occupied most of the spreading time, has a relation of R–$t^{0.251}$ obtained from the MD results. The spreading exponent of 0.251 is close to 1/5 predicted by molecular kinetic theory (MKT) [29]. The results indicate that nanodroplet spreading follows kinetic theory in which molecular absorption–desorption dominates the spreading, rather than the hydrodynamic models in which the gravity/capillary forces drive the spreading. In addition, the fitted friction coefficient of pure water as shown in Fig. 3.3b consists of the experimental results [30]. The details have been discussed in Chap. 3.

6.3 Results and Discussion

6.3.1 Effects of Substrate Temperatures

Figure 6.3 shows the snapshots ($t = 6$ ns) of the spreading–evaporating nanofluid droplet on the gold substrates at various temperatures. The solid-like ordering

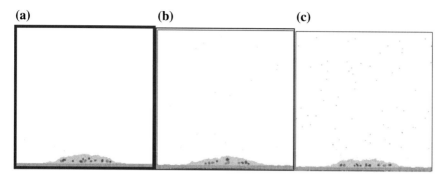

Fig. 6.3 Snapshots of nanofluid droplets spreading with evaporating at various solid surface temperatures: (t = 6 ns): **a** T_s = 300 K; **b** T_s = 333 K; **c** T_s = 363 K

Fig. 6.4 Spreading radii for nanofluid droplets at various solid surface temperatures (φ = 6.77 %, d_0 = 10.0 nm)

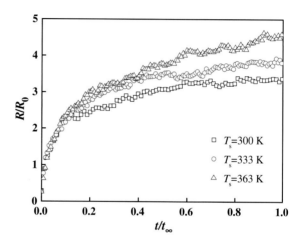

nanoparticle structure was also not observed in present nanometer droplet. Unlike the spreading without heating conditions in Chap. 3, the nanoparticles deposited uniformly at the bottom of the droplet. The deposited particles during the droplet spreading with evaporating are related to the phenomena of the "coffee ring," which had been widely studied for the micron or millimeter colloidal suspension [31–35]. Shen has reviewed the topic of "coffee ring" and also provided a guideline to suspend the coffee ring effects [36]. The trick is to reduce the size of spreading droplet to the submicron or nanometer, which consists of the results of present simulations. The present results provide the potential techniques to control the self-assembly of nanoparticles by tuning the external thermal fields or others.

As discussed in the previews chapters, the nanoparticle self-assembly behaviors affect the dynamic wetting by nanofluids. Therefore, the uniform nanoparticle deposition will lead to different spreading behaviors compared with the spreading without external thermal fields, which is shown in Fig. 6.4. The increasing substrate temperature greatly increases the wetted area of the droplet on the gold substrate. For T_s = 300 K, the droplet reached the equilibrium size for $t > 2$ ns. The evaporation also greatly affects the spreading areas. For the limited evaporating condition ($T_s = 300$ K and $T_l = 300$ K), the equilibrium wetting radius is 15.5 ± 0.2 nm but 19.2 nm for $T_s = 333$ K. The spreading radius for $T_s = 363$ K is greater than 23.4 nm since equilibrium was not yet reached at $t = 6$ ns. The contact line motion is driven by the resultant force from the surface tension force and the adhesion force between the solid and liquid phases and overcomes the resistant force due to viscosity. Therefore, the results can also be explained from the driven force of contact line motion, the unbalanced Young's stress, F, which has been discussed repeatedly in previous chapters.

The role of evaporation on the spreading was studied by calculating the surface tension forces between the solid–liquid and the liquid–vapor. The solid–liquid surface tension is difficult to measure experimentally. The solid–liquid surface tension was calculated from MD simulations using three NVT systems, one with gold substrate wetted by a water film layer and the others with only the bulk liquid or bulk gold. The time-averaged excess energy of these three systems ($\langle E_{g-w} \rangle - \langle E_w \rangle - \langle E_g \rangle$) was then used to calculate the surface tension ($\langle E_{g-w} \rangle - \langle E_w \rangle - \langle E_g \rangle)/A_{\text{interface}}$ for gold–water, σ_{SL}. For the three substrate temperatures, the solid–liquid surface tension is 0.514 ± 0.03 N $\cdot$ m^{-1}, indicating that σ_{SL} is independent of temperature. The surface tensions of many metals (e.g., Ag, Au, and Cu) are much larger than those of pure water (1.145 N $\cdot$ m^{-1} for gold at 1065 °C) and decrease with decreasing temperature at a rate of only 0.0002 N $\cdot$ m^{-1} $\cdot$ K^{-1} [37], which means that the σ_{SL} is only a weak function of temperature. Therefore, the variation of liquid–vapor surface tension, σ_{LV}, is the main reason for the changes in the spreading–evaporating process as temperature increases. The liquid–vapor surface tension was determined using simulations of two NVT systems, one with two flat-free surfaces and the other with only the bulk liquid, as shown in Fig. 6.5. The simulation details have been discussed in Chaps. 3 and 4 with a surface tension of 0.06871 N/m at $T = 300$ K for pure water, indicating the validity of the calculation [38]. The liquid–vapor surface tension decreases with increasing substrate temperature, as shown in Table 6.2, leading to a greater contact line motion driven force, a faster contact line spreading velocity, and a large spreading area at the equilibrium stage. The results also indicate that the nanofluid dynamic wetting with evaporating in the present work is dominated by the modified surface tension, rather than the structural disjoining pressure since there is no nanoparticle self-assembly near the contact line region. The external thermal conditions change the nanoparticle sedimentation behaviors and hence affect the dynamic wetting.

6.3.2   Microscopic Pictures of Spreading–Evaporating Nanodroplets

The contact line motion in molecular kinetic theory involves a process of adsorption and desorption of molecules to and from the substrate. Figure 6.6 shows the nanoparticle motion and water molecule movement in the vicinity of the contact

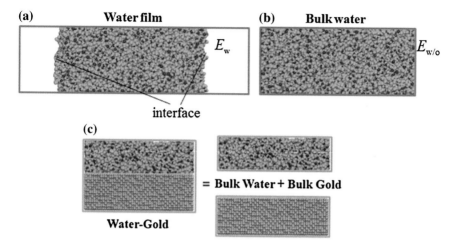

Fig. 6.5 Schematic simulation method of liquid–vapor surface tension (**a** and **b**) and solid–liquid interface tension (**c**). Reprinted from Ref. [28], with kind permission from ASME

Table 6.2 Water surface tension for various temperatures

Temperature (K)	MD (N/m)	Experiment (N/m)	$\varphi = 6.77$ % Nanofluids (N/m)
300	0.06871	0.07197	0.1109
333	0.06262	0.06610	0.1026
363	0.05840	0.06066	0.0913

Reprinted from Ref. [28], with kind permission from ASME

line region (3 nm from the contact line) between 3 and 4 ns. 531 water molecules were dyed and highlighted at $t = 3$ ns for $T_s = 300$ K, with 483 dyed for $T_s = 333$ K and 398 dyed for $T_s = 363$ K. Most of the water molecules move forward along with the contact line at $T_s = 300$ K. The nanoparticles cannot move into the contact line regions when the contact line moving forward. Therefore, the solid-like ordering structures due to the nanoparticle self-assembly cannot be observed here. As the substrate temperature increases, more dyed molecules are transported backward into the bulk droplet. Figure 6.7 shows the replacement portion, which is the ratio of dyed water molecules to the total water molecules in the new wetted region. More than 60 % of water molecules in the newly wetted region come from the initially dyed molecules at 3 ns during droplet spreading for $T_s = 300$ K, while more than 50 % for $T_s = 333$ K and 40 % for $T_s = 363$ K. However, the molecular mobility becomes more violent as temperature increases. A larger portion of the undyed molecules in the bulk droplet transports to the newly wetted region as temperature increases, with about 40 % for $T_s = 300$ K, 50 % for $T_s = 333$ K, and 60 % for $T_s = 363$ K (corresponding to 126 molecules for $T_s = 300$ K, 174 molecules for $T_s = 333$ K, and 216 molecules for $T_s = 363$ K), which facilitates droplet spreading at high temperatures.

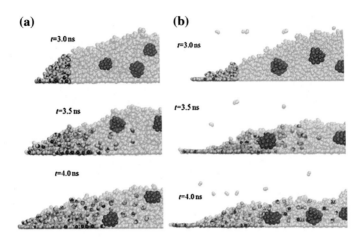

Fig. 6.6 The molecular mobility trajectory at various solid surface temperatures: **a** without evaporation (T_s = 300 K, T_L = 300 K); **b** with evaporation (T_s = 363 K, T_L = 300 K)

Fig. 6.7 Water molecular portions from bulk liquid to the new wetted area for nanofluid droplets (φ = 6.77 %, d_0 = 10.0 nm, t = 3–4 ns) spreading at various surface temperatures

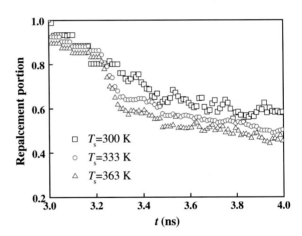

6.3.3   Effect of Initial Droplet Temperature

The nanofluid droplet spreading was further analyzed with three different initial temperatures (300, 333, and 363 K) on the 363 K substrate. The evaporation rate for T_l = 363 K is 5.0 % at 1 ns, 2.3 % for T_l = 333 K, and 0 % for T_l = 300 K. The growth rates for the normalized spreading radius are shown in Fig. 6.8 for the three droplet temperatures. The results show that the initial droplet temperature has very limited effect on the wetting kinetics. Figure 6.9 shows the droplet temperature evolution. When the droplet with d_0 = 10 nm touched the heated gold substrate, the droplet temperature increased sharply and equilibrated with substrate temperature of 363 K at t = 0.26 ns for T_l = 300 K and t = 0.092 ns for T_l = 333 K, which is much shorter than the entire spreading time of 6 ns, accounting for only 3.3 % spreading

Fig. 6.8 Effects of nanofluid droplet ($\varphi = 6.77$ %, $d_0 = 10$ nm) initial temperatures on the spreading–evaporating process ($T_s = 363$ K)

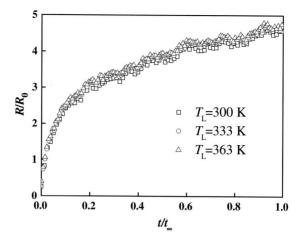

Fig. 6.9 The variation of droplet temperatures ($T_s = 363$ K): **a** 10-nm nanofluid droplet; **b** 13.5-nm fluid droplet

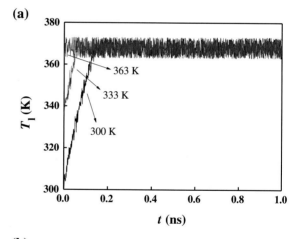

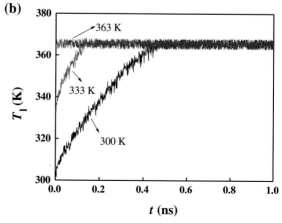

time for $T_1 = 300$ K and 1.2 % for $T_1 = 333$ K. The very rapid increase of the droplet temperature leads to very small differences in radius versus time curves for the various initial droplet temperatures. This trend was still observed for a larger droplet (9000 water molecules, with diameter of 13.5 nm) as shown in Fig. 6.9b. The droplet temperature increase is relatively slower for the larger droplet, but still fast with a heating duration of 0.82 ns for $T_1 = 300$ K, accounting for 3.8 % of the total spreading time (10 ns for the 9000 molecules droplet), and 0.036 ns (1.3 %) for $T_1 = 333$ K. Thus, for spreading–evaporating nanodroplet, the heating is much faster than the spreading. In addition, the additional nanoparticles increase the heat exchange rates. Therefore, the initial droplet temperatures have very little effect on the wetting kinetics of nanodroplets.

Fig. 6.10 Effects of solid surface wettability on the spreading–evaporation process: **a** without evaporating ($T_L = 300$ K, $T_s = 300$ K); **b** with evaporating ($T_L = 300$ K, $T_s = 363$ K)

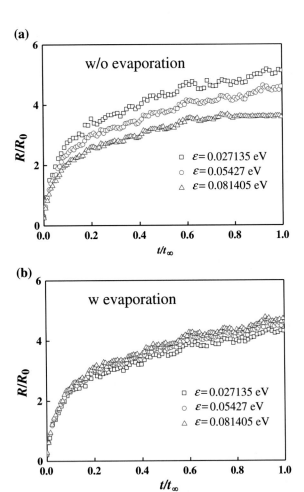

6.3.4   Effects of Wettability

The effects of substrate wettability on the droplet wetting kinetics are quite different with and without evaporation as shown in Fig. 6.10. The substrate wettability was modified by changing the interaction parameter, ε, between the gold atoms of substrates and the water molecules. As ε increased, the interaction force between substrate and the water molecules increased, leading to a hydrophilic substrate. For the limited evaporating case (w/o) in Fig. 6.10a, the wettability strongly affects the spreading due to the stronger gold–water interaction as ε increases. However, for the evaporating case in Fig. 6.10b ($T_s = 363$ K, $T_l = 300$ K), the water molecules become very active and are transported violently and randomly in the bulk droplet, so the attraction to the substrate becomes relatively insignificant, leading to a very limited effects of the gold–water interactions on the wetting kinetics for the spreading–evaporating case. The absorption–desorption process in the vicinity of the contact line region becomes more frequent and intensive.

6.4   Conclusions

The wetting kinetics of nanofluid nanodroplets with external heating conditions were simulated using molecular dynamic simulations. The effects of the initial droplet temperature, substrate temperature, and wettability were examined.

1. The molecule and nanoparticle random motion become significant with the heating conditions, accelerating the nanoparticle diffusion. The nanoparticle self-assembly near the contact line region was not observed. Instead, the nanoparticles tend to deposit uniformly at the bottom of droplet on the solid surface. The nanoparticle sedimentation rate increases with increasing temperature.
2. The reduction of the liquid–vapor surface tension due to the increasing of substrate temperatures strongly affects the contact line movement for the nanofluid droplet. The effects of evaporation on wetting kinetics are enhanced by the frequency and intensity of water molecule absorption–desorption in the vicinity of contact line region.
3. The initial droplet temperature had little effect on the contact line motion due to the fast heating process of the nanometer droplet spreading. The additional nanoparticles enhance the heat exchange between the substrates and the droplet molecules. The substrate wettability affected the spreading much more for spreading with limited evaporation than spreading with intense evaporation.

References

1. deGennes PG (1985) Wetting: statics and dynamics. Rev ModPhys 57:827–863
2. Bonn D, Eggers J, Indekeu J et al (2009) Wetting and spreading. Rev Mod Phys 81:739–805
3. Oron A, Davis SH, Bankoff SG (1997) Long-scale evolution of thin liquid films. Rev Mod Phys 69:931–980
4. Birdi KS, Vu DT (1988) A study of the evaporation rates of small water drops placed on a solid surface. J Phys Chem 93:3702–3703
5. Sadhal SS, Plesset MS (1979) Effect of solid properties and contact-angle in dropwise condensation and evaporation. ASME J Heat Transfer 101:48–54
6. Deegan RD, Bakajin O, Dupont TF et al (2000) Contact line deposit in an evaporating drop. Phys Rev E 62:756–765
7. Hu H, Larson RG (2002) Evaporation of a sessile droplet on a substrate. J Phys Chem B 106:1334–1344
8. Bourges-Monnier C, Shanahan MER (1995) Influence of evaporation on contact angle. Langmuir 11:2820–2829
9. Neumann AW, Tanner W (1970) The temperature dependence of contact angles-ploytetrafluoroethylene/n-decane. J Colloid Interface Sci 34:1–8
10. Galliker P, Schneider J, Eghlidi H et al (2012) Direct printing of nanostructures by electrostatic autofocussing of ink nanodroplets. Nat Commun 3:890
11. Wang XD, Lee DJ, Peng XF et al (2007) Spreading dynamics and dynamic contact angle of non-Newtonian fluids. Langmuir 23:8042–8047
12. Liang ZP, Wang XD, Duan YY et al (2012) Energy-based model for capillary spreading of power-law liquids on a horizontal plane. Colloid Surface A 403:155–163
13. Joanny JF (1986) Dynamics of wetting—interface profile of a spreading liquid. Journal de Mecanique Theoriqueet Appliquee, SI, pp 249–271
14. Lu G, Duan YY, Wang XD et al (2011) Internal flow inevaporating droplet on heated solid surface. Int J Heat Mass Transfer 54:4437–4447
15. Bhardwaj R, Fang XH, Attinger D (2009) Pattern formation during the evaporation of a colloidal nanoliter drop: a numerical and experimental study. New J Phys 11:1–33
16. Strotos G, Gavaises M, Theodorakakos A et al (2008) Numerical investigation on the evaporation of droplets depositing on heated surfaces at low Weber numbers. Int J Heat Mass Transfer 51:1516–1529
17. Shih CH, Wu CL, Chang LC et al (2011) Lattice Boltzmann simulations of incompressible liquid–gas systems on partial wetting surfaces. Philos Trans R Soc London Ser A 369:2510–2518
18. Yan YY, Zu YQ (2007) Lattice Boltzmann method for incompressible two-phase flows on partial wetting surface with large density ratio. J Comput Phys 227:763–775
19. Attar E, Koerner C (2009) Lattice Boltzmann method for dynamic wetting problems. J Colloid Interface Sci 335:84–93
20. Coninck JD, D'Ortona U, Koplik J et al (1995) Terraced spreading of chain molecules via molecular dynamics. Phys Rev Lett 74:928–931
21. Khan S, Singh JK (2013) Wetting transition of nanodroplets of water on textured surfaces: a molecular dynamics study. Mol Simulat 40:458–468
22. Yang JX, Koplik J, Banavar JR (1992) Terraced spreading of simple liquids on solid surfaces. Phys Rev A 46:7738–7749
23. Coninck JD, Blake TD (2008) Wetting and molecular dynamics simulations of simple liquids. Annu Rev Mat Res 38:1–22
24. Huh C, Scriven LE (1971) Hydrodynamic model of steady movement of a solid–liquid–fluid apparent contact line. J Colloid Interface Sci 35:85–101
25. Zhang JG, Leroy F, Muller-Plathe F (2013) Evaporation of nanodroplets on heated substrates: a molecular dynamics simulation study. Langmuir 29:9770–9782

26. Yang X, Yan YY (2011) Molecular dynamics simulation for microscope insight of water evaporation on a heated magnesium surface. Appl Therm Eng 31:640–648
27. Hwang CC, Lee SC, Hsieh JY (1999) A study of temperature effects on drop spreading by the molecular dynamics simulation. J Phys Soc Jpn 68:3742–3743
28. Lu G, Duan YY, Wang XD (2015) Effects of free surface evaporation on water nanodroplet wetting kinetics: a molecular dynamics study. J Heat Transfer 137:091001–091001–6
29. Blake TD, Haynes JM (1969) Kinetics of liquid/liquid displacement. J Colloid Interface Sci 30:421–423
30. Blake TD (2006) The physics of moving wetting lines. J Colloid Interface Sci 299:1–13
31. Wong TS, Chen TH, Shen XY et al (2011) Nano-chromatography driven by the coffee ring effect. Anal Chem 83:1871–1873
32. Deegan RD, Bakajin O, Dupont DF, Witten TA et al (1997) Capillary flow as the cause of ring stains from dried liquid drops. Nature 389:827–829
33. Hu H, Ronald L (2006) Marangoni effect reverses coffee-ring depositions. J Phys Chem B 110:7090–7094
34. Weon BM, Je JH (2010) Capillary force repels coffee-ring effect. Phys Rev E 82:015305
35. Yunker PJ, Still T, Lohr MA et al (2010) Suppression of the coffee-ring effect by shape-dependent capillary interactions. Nature 476:308–311
36. Shen XY, Ho CM, Wong TS (2010) Minimal size of coffee ring structure. J Phys Chem B 114:5269–5274
37. Adam NK (1941) The physics and chemistry of surfaces. Oxford University Press, London
38. Dean JA (1961) Lange's handbook of chemistry, 10th edn. McGraw Hill, New York

Chapter 7
Conclusions and Prospects

Abstract In this chapter, we provide the general conclusions and the contributions of this book. The prospects in dynamic wetting by nanofluids are also provided in this chapter.

7.1 Conclusions

Due to their special flow and thermal properties, nanofluids were widely used in heat transfer enhancement and drag reduction devices, as well as in the drug delivery and micro/biofluid systems, in which the dynamic wetting by nanofluids plays significant roles. The mechanism of dynamic wetting by nanofluids is still unclear due to limitations of nanoscale experimental techniques and fundamental theories. Studies of the dynamic wetting by nanofluids face great challenges since the wetting behavior crosses several length and timescales. This book analyzes the effects of the bulk and local dissipation in the nanofluids due to the transport and self-assembly of nanoparticles on the macroscopic dynamic wetting behavior using macroscopic experiments and multiscale simulation methods. The results describe both the macroscopic and microscopic mechanisms and tunable methods to control nanofluid dynamic wetting. The main conclusions are as follows:

1. The time-dependent wetting radius and contact angle for various dilute nanofluid droplets were measured by the droplet spreading method. The effects of the nanoparticle material, loading and diameter, the base fluid, and the substrate material were examined. The results show that the adding of nanoparticles inhibits the dynamic wetting of nanofluids as compared with base fluids. The reduced spreading rate can be attributed to the increase in either surface tension or viscosity due to adding nanoparticles into the base fluid. It is interesting that once the effects of the surface tension and viscosity are both eliminated using the non-dimensional analysis, the wetting radius versus spreading time curves for all the nanofluid droplets overlap each other. In addition, the spreading exponent fitted from the nanofluid dynamic wetting data is found to be very close to 0.1,

© Springer-Verlag Berlin Heidelberg 2016
G. Lu, *Dynamic Wetting by Nanofluids*,
Springer Theses, DOI 10.1007/978-3-662-48765-5_7

which meets the prediction of the classical hydrodynamic model derived from the bulk viscous dissipation approach. Thus, the present results prove that the spreading of the dilute nanofluid droplets is dominated by the bulk dissipation rather than by the local dissipation at the moving contact line.

2. The dynamic spreading of water nanodroplets containing non-surfactant nanoparticles and the effects of structural disjoining pressure are examined via molecular dynamic simulations. The nanoparticle diffusion time is larger than the nanosize droplet spreading time. The nanoparticles do not have enough time to diffuse to the vicinity of contact line region; thus, the self-assembly of nanoparticles does not occur. The addition of non-surfactant nanoparticles hinders rather than enhances the droplet spreading kinetics during the nanosecond process. The contact line velocity decreases with increasing nanoparticle volume fraction and particle–water interactions, as a result of increasing surface tension and solid–liquid friction and the absence of nanoparticle ordering in the vicinity of contact line. The structural disjoining pressure is ten times larger than the unbalanced Young's stress, which can facilitate the contact line motion if the nanoparticle self-assembly occurs in the vicinity of contact line region.

3. The surface tension, viscosity, and rheology of gold–water nanofluids were calculated using molecular dynamic simulations which provide a microscopic interpretation for the modified properties on the molecular level. The gold–water interaction potential parameters were changed to mimic various nanoparticle types. The results show that the nanoparticle wettability is responsible for the modified surface tension. Hydrophobic nanoparticles always tend to stay on the free surface, so they behave like a surfactant to reduce the surface tension. Hydrophilic nanoparticles immersed in the bulk fluid impose strong attractive forces on the water molecules at the free surface which reduce the free surface thickness and increase the surface tension of the nanofluid. Solid-like absorbed water layers were observed around the nanoparticles which increase the equivalent nanoparticle radius and reduce the mobility of the nanoparticles within the base fluid which increase the nanofluid viscosity. The results show the water molecule solidification between two or many nanoparticles at high nanoparticle loadings, but the solidification effect is suppressed for shear rates greater than a critical shear rate; thus, Newtonian nanofluids can present shear-thinning non-Newtonian behavior.

4. A mesoscopic study of the nanofluid wetting kinetics using the lattice Boltzmann method was conducted to investigate the effects of nanoparticle motions in nanoscale (10^{-9} m) on the dynamic wetting behaviors that occur in the macroscopic scale (10^{-3} m). The effects of nanoparticle motion in the bulk liquid contribute to the bulk dissipations, which modify the surface tension and the rheology of based fluids. The self-assembly of nanoparticles in the vicinity of the contact line regions contributes to the local dissipations. The effects of nanoparticle bulk and local dissipations on the macroscopic dynamic wetting were studied using mesoscopic simulations. The results show that the LBM is capable to simulate both the microscopic phenomena and the macroscopic

dynamic wetting. Adding hydrophobic nanoparticles facilitates the dynamic wetting, while adding hydrophilic nanoparticles deteriorates the dynamic wetting. The shear-thinning non-Newtonian behavior due to the addition of nanoparticle enhances the dynamic wetting of nanofluids. For the partial wetting droplet, the structural disjoining pressure due to the self-assembly of nanoparticle in the vicinity of contact region enhances the contact line motion. The nanoparticle global deposition has few effects on the dynamic wetting. The study provides multiscale understanding and tunable methods of the nanofluid dynamic wetting.

5. The wetting kinetics of a nanofluid nanodroplet with evaporation on a heated gold substrate were simulated using molecular dynamic simulations. The effects of the initial droplet temperature, substrate temperature, and wettability were examined. The molecule and nanoparticle random motion become significant with the heating conditions, accelerating the nanoparticle diffusion. The nanoparticle sedimentation rate increases with increasing temperature. The nanoparticles tend to deposit uniformly at the bottom of the droplet on the solid surface. The reduction of the liquid–vapor surface tension due to the increasing of substrate temperatures strongly affects the contact line movement for the nanometer water droplet. The effects of evaporation on wetting kinetics are enhanced by the frequency and intensity of water molecule absorption–desorption in the vicinity of contact line region. The initial water droplet temperature had little effect on the contact line motion due to the fast heating process for the nanometer droplet spreading. The substrate wettability affected the spreading much more for spreading with limited evaporation than for spreading with intense evaporation.

7.2 Prospects

As mentioned in Chap. 1, the study of nanofluid dynamic wetting still faces great challenges due to the lack of nanoscale and multiscale experimental techniques and theories. There is still a long way to go in the future before solving these challenges. The present study only draws the branches and trunk in the study of dynamic wetting by nanofluids, and more leaves and flowers are needed to make the trees luxuriant. The prospects of dynamic wetting by nanofluids lie in several aspects, including the nanoscale experimental evidences of nanoparticle self-assembly near the contact line region, theoretical model with combination of the effects of the bulk and local dissipations, and the numerical model with the capacity of simulating sufficient nanoparticle number. One candidate of experimental techniques to detect the self-assembly of nanoparticles near the contact line region is atomic force microscopy (AFM), which has been reported in the detection of nanoscale profile of contact line region. The hybrid of the molecular kinetic theory and the hydrodynamic model could be the potential option to establish the dynamic wetting theoretical model of nanofluids, which can combine the bulk and local dissipations revealed in this study. In the numerical studies, the lattice Boltzmann method is the promising tools in the simulation of dynamic wetting by nanofluids. However, a reasonable colloidal model is still needed in simulating the nanofluids.

Printed in the United States
By Bookmasters